Lecture Notes in Economics and Mathematical Systems

464

Springer
Berlin
Heidelberg
New York
Barcelona
Budapest
Hong Kong
London
Milan
Paris
Singapore
Tokyo

Arvid Aulin

The Impact of Science on Economic Growth and its Cycles

The Mathematical Dynamics Determined by the Basic Macroeconomic Facts

Springer

Author

Prof. Arvid Aulin
The Finnish Academy of Science and Letters
Oulunkyläntori 2 C 16
00640 Helsinki, Finland
Fax: 358-9-7283042
Email: aulin@nettilinja.fi

Cataloging-in-Publication Data applied for

Die Deutsche Bibliothek - CIP-Einheitsaufnahme

Aulin, Arvid:
The impact of science on economic growth and its cycles : the mathematical dynamics determined by the basic macroeconomic facts / Arvid Aulin. - Berlin ; Heidelberg ; New York ; Barcelona ; Budapest ; Hong Kong ; London ; Milan ; Paris ; Santa Clara ; Singapore ; Tokyo : Springer, 1998
(Lecture notes in economics and mathematical systems ; 464)
ISBN-13: 978-3-540-64727-0 e-ISBN-13: 978-3-642-95861-8
DOI: 10.1007/978-3-642-95861-8

ISSN 0075-8442
ISBN-13: 978-3-540-64727-0

Typesetting: Camera ready by author
SPIN: 10546692 42/3143-543210 - Printed on acid-free paper

Contents

1. THE MACROECONOMIC PROBLEM 1

1.1. The Basic Macroeconomic Facts 2
1.2. This Theory Fits Well the Facts But Does It Fit the Current Dogmas? 5
1.3. Why This Dynamics Explains the Facts While the Current Dogmas Fail So Badly? 7
Reason No.1: Generalized human capital 7
Reason No.2: Generalized value function 10

Part 1
THE FUNDAMENTAL LAWS OF SCIENCE-BASED ECONOMIC GROWTH

2. CANONICAL FORMALISM: BASICS 15

2.1. Elementary Mathematics 15
Hamiltonian formalism 15
The origins of canonical formalism 17
2.2. The Use of Canonical Formalism in Economics 22
Maximization of the value function 22
Example: The Solow model 23
The Arrow-Kurz generalization 28
Example: The Lucasian mechanics of economic growth 30

3. THE CANONICAL FORMALISM OF MACROECONOMICS 35

3.1. Generalized Value Function 35
The uses of human time 35

Scientific avantgarde of technological progress 37
The pursuit of scientific progress and innovations expands radically the macroeconomic value function 38
3.2. Growth Cycles 41
Cycle equations: The first normal form 41
Fact 9 and the continuous invariance group 43
Cycle equations: The second normal form 45
The ordinary cycles and the anomalous cycles 46
Linear approximations 48
3.3. Solving the Fundamental Equations 51
General solution algorithm 51
Positive growth rate of human capital over cycles 52
Boundary conditions 55
3.4. Cycle Functions Generating the Business Cycles 57
Cycle function of the output Y 57
Cycle function of the consumption C 58
Cycle function of the investment I 59
Cycle function of the human capital h 60
Cycle function of the employment E 60
Cycle function of the productivity Y/E 61
The cycle functions as causal replacements of Slutsky statistics in business cycle theory 61

Part 2
DERIVATION OF THE 17 BASIC MACROECONOMIC FACTS FROM THE CANONICAL FORMALISM

4. THE PRINCIPLE OF ECONOMY IN SCIENTIFIC EXPLANATION 67

Immediate effects of the generalized value function 68
The prediction power of cycle functions 69

5. GROWTH PATHS DETERMINED BY CANONICAL FORMALISM 72

5.1. The Ordinary Growth Path: Balanced Growth 72
Solution algorithm 72
The Kaldor facts derived from theory 74
The boundary conditions 75
5.2. Anomalous Growth Path After a Crash 77
Solution algorithm: Moving cycle center 77
Verification by the Solow (1957) Statistics 80

6. THE PLOSSER FACT AND OTHER GROWTH EFFECTS OF SAVINGS RATE 85

6.1. The Extended Growth Effects of Savings Rate Following from the Extended Value Function 85
Twofold dependence on savings rate 86
The differential method 87
6.2. Empirical Verifications of the Extended Growth Effects of Savings Rate in Various Countries 89
Problems of economic statistics 89
Growth effects of savings rate in the U.S.A. 91
Growth effects of savings rate in Japan 93
Growth effects of savings rate in Canada 96
Growth effects of savings rate in France 99
Conclusions 101
Slowdown of productivity and savings rate: a general perspective 102

7. THE KYDLAND FACTS ON LEADS AND LAGS IN BUSINESS CYCLES 105

7.1. The Importance of Labour Input and the Productivity of Labour 105
7.2. The Maximum and Minimum Points in Detrended Business Cycles 108
7.3. The Leads and Lags in the Second Half of the Century 114

Leads and lags in the Golden Age period 1947-73 114
Leads and lags in the period 1950-81 118
7.4. The Leads and Lags in the First Half of the Century 121
Conclusions from the analysis of leads and lags 124

8. Correlations and Variances over the Ordinary Business Cycles 125

8.1. Calculation of Theoretical Correlations and Standard Deviations 125
8.2. Comparisons with Facts and the Stochastic Models of Kydland-Prescott, Hansen-Rogerson and Danthine-Donaldson 130
8.3. Comparisons with Empirical Autocorrelations 134
8.4. Which Predicts Better the Data: The Latest Standard Stochastic Model (Cooley-Prescott) or the Minimal Macroeconomic Dynamics? 138
8.5. Business Cycles of Total Working Hours 141
The cycle function of total hours defined 141
Calculation of output correlation and SD 142
Predictions compared with facts 143
8.6. Comments and Challenges 146
Relative independence of predictions from calibration 146
Two levels of macroeconomic theory 146
Convergent instead of divergent cycle functions? 148

9. Correlations over Anomalous Business Cycles 149

9.1. Why Anomalies Are Important in Science? 149
Approximation of anomalous cycle functions 151
Integrations over the period of a cycle 153
9.2. Theoretical Explanation of Observed Anomalies in the Business Cycle 156
Reduced procyclicality of consumption: U.S.A., U.K. 156
Retained procyclicality of employment: U.S.A., U.K. 158
Radically reduced procyclicality of investment: U.S.A. 160
Productivity turned anticyclical: U.K. 163

Causal background 166
APPENDIX (CHAPTER 9) 167

10. THE ROLE OF STOCHASTIC SHOCKS IN THE BUSINESS CYCLE 172

10.1. How Important are Stochastic Shocks? 172
10.2. Construction of Stochastic Cycle Functions 174
Random series with a definite mean and SD 174
Stochastic cycle functions and their effects on correlations and standard deviations 174
10.3. Numerical Calculations and Conclusions 178
Calibration 179
Small but not negligible effects of shocks 180
Final comments 182
APPENDIX (CHAPTER 10) 186
REFERRED LITERATURE 196
INDEX 199

1 The Macroeconomic Problem

The world of observations uniquely determines the theoretical system [in mathematical sciences] ... but there is no logical way leading to the fundamental laws, only intuition.
(Einstein in his speech in honor to Max Planck in 1918.)

Does it apply to economics? In mathematical natural sciences, such as physics and chemistry, theoretical development has indeed been largely determined by observed facts. In economics the situation is a little different. Economics has become an influential practical science, like medicin and law. It is like medicin and law a *profession*, where professional solidarity plays an important role. Being a profession and a science simultaneously implies that theorizing in economics is partly determined by the general wisdom of the day, as accepted in the profession. Because of this important professional connection the basic theoretical assumptions at any given time tend to be widely accepted over most of the profession. It follows that new fundamental ideas have come slowly in economics, where theoretical foundations carry that professional responsability.

For this and other reasons the formation of theories in economics has usually followed another line. The dominant wisdom in the economist profession tends to prevail over a rather long period. But in macroeconomics especially it has changed several times, and the fundamental changes when they have come have not been generalizations like in physics (from Newton to Einstein and to quantum theory) but often rather abrupt replacements of basic ideas, after having been preceded by a long period during which theoretical attitudes were unchanged.

Sometimes macroeconomic theory has been allowed to stray far from the empirics. A case in point is growth theory, which could live and proceed a long time without much connection with empirical observations. Lately the theory of endogeneous growth[1] has brought it closer

[1]See e.g. Paul M.Romer, The Origins of endogeneous growth, Journal of Economic Perspectives 8,1994, p.3-22.

to the empirics. But still a wide gap exists between theory and facts, even in the most promising new theoretical approaches.[2]

In this book just the opposite is made, proving that the Einstein principle works also in macroeconomics. But this implies the introduction of a new mathematical element to economic theory, viz.the concept of *group-theoretical invariance*. It proves that this element is necessary for the explanation of the basic macroeconomic facts. It is also necessary for a correct representation of growth cycles with irregularly varying lengths. The corrected macroeconomic dynamics is constructed (Chapter 3) and empirically verified (Part 2).

1.1 The Basic Macroeconomic Facts

Concerning average growth:

Most empirical statistics reveal an average growth that is approximately covered by a balanced-growth path on which the six stylized facts, listed by Kaldor (1963) and reproduced by Barro and Sala-i-Martin[3], and a seventh one emphasized by Plosser[4] are valid:

1. Per capita output grows over time, and its growth rate does not tend to diminish.
2. Physical capital per worker grows over time.
3. The rate of return to capital is nearly constant.
4. The ratio of output to physical capital is nearly constant.
5. The shares of labour and physical capital in national income are nearly constant.
6. The growth rate of output per worker differs substantially across countries.
7. The growth rate of output per worker (i.e. productivity of labour) depends positively on the rate of investment, while such a law does not hold true between the rate of investment and the growth rates of output and employment.

---

[2]An example of which is George Evans, Seppo Honkapohja and Paul Romer, Growth Cycles (manuscript dated September 26, 1996: see the home page of Paul M. Romer).

[3]Robert J.Barro and Xavier Sala-i-Martin, Economic Growth, McGraw-Hill New York 1995, p.5.

[4]C.I.Plosser, The search for growth, Federal Reserve Bank of Kansas City Economic Review, Symposium, 1992, p.57-86.

Note 1. The Kaldor facts are trivial and are reproduced on the balanced-growth path of many theories including this one (Section 5.1). The Plosser fact 7 is nontrivial and its first exact and quantitatively amply verified explanation is given in this book (Chapter 6).

Concerning irregular growth variations:

8. The existence of growth variations in the form of irregular shocks.

9. The existence of growth cycles called 'business cycles', with irregularly varying lengths having an average of 4 years in the U.S.A.

Note 2. Two mathematical methods are available for a potential explanation of these irregularities:

(a) The fact 8 suggests the introduction of stochastic processes generating the shocks and the cycles. This is the easy solution, but it seems unable to explain the facts 7,10-12 or 16-17, and it is weak also on the facts 13-15. The Einstein principle thus excludes this method.

(b) The fact 9 suggests as an alternative solution cycle equations with an invariance group involving time (for details see p.43-44). The dynamics constructed in this book gives this solution. The shocks are superposed on cycles but prove to have a minor role.

The method (a) dominates today in economic theory, but only the method (b) is in agreement with all the basic macroeconomic facts.

Concerning quantitative regularities of growth cycles:

The quantitative facts discussed by Kydland[5] concerning the leads and lags of aggregate economic variables with respect to output:

10. The productivity of labour has mostly a lead of 2-3 quarters, while a lead of one or four quarters is still possible.

11. Employment has a lag of one to one and a half quarters.

12. Total hours has not any remarkable lead or lag.

Note 3. The facts 10-12 are nontrivial and are well reproduced by the dynamics of this book (Chapter 7), while the conventional RBC-models following the method (a) fail completely.

[5]For these facts see Table 3 on p.106 of this book reproducing part of Table 5.1 in: Finn Kydland, Business cycles and aggregate labor market fluctuations, in Thomas F.Cooley (ed.), Frontiers of Business Cycle Research, Princeton University Press 1995.

The empirical growth cycles, when averaged over a long enough period of time, also display the following correlations of aggregate economic variables with output, the following standard deviations, and the following autocorrelations of output, over the detrended cycle:

13. Investment, consumption, employment and total hours have a high output correlation (with observations varying from .70 to .95), productivity of labour a small one (from .10 to .40).

14. The standard deviation (SD) of investment is 3-5 times that of output, the SD of total hours has been measured to be about .95 to 1.10 times that of output, the SD of employment about .60 to .80 times that of output, the SD of consumption and that of productivity about .35 to .70 times the SD of output.

15. The autocorrelations of output must be close to those given by Kydland and Prescott first in 1982[6] and improved in 1990[7].

Note 4. It will be shown in Chapter 8 that the predictions of the dynamics constructed in this book are much closer to the facts 13-15 than are the predictions obtained from the conventional RBC-models. Here again the predictions obtained by the method (b) are superior to those given by the method (a). The numbers refer to U.S. economy.

To these well-known 15 fundamental facts the following less generally known *deviations* from them can be added.

Anomalities of growth and growth cycles:

16. In a deviation from the stylized facts of Kaldor, the ratio of output to physical capital left the balanced-growth path in the Great Depression, and followed in the U.S.A. during the period 1930-48 a growth path on which this ratio grew following a logistically rising path. This deviation was registered by the 1957 statistics of Solow[8], which is reprinted here in Table 1. Since Solow discussed another problem he ignored this deviation, which however is faithfully reproduced and

[6]Kydland,F. and E.Prescott, Time to build and aggregate fluctuations, Econometrica 50,1982, p.1345-1370.

[7]Kydland,F. and E.C.Prescott, Business cycles: real facts and monetary myth, Federal Reserve Bank of Minneapolis Quarterly Review 14, 1990, p.3-18.

[8]R.M.Solow, Technical change and the aggregate production function, The Review of Economics and Statistics, 312-320, Table 1.

quantitatively approximated (Section 5.2) by the dynamics constructed in this book, using the method (b). The conventional stochastic method (a) cannot explain this anomaly at all, which is perhaps one of the reasons why the fact 16 has been entirely ignored in economics.

17. In the period (1914-50) covering the deviation period 1930-48 in the U.S.A. also a corresponding deviation has been observed in the correlations of aggregate economic variables with output over the detrended cycle. This deviation was registered much later by Correia, Neves and Rebelo[9]. They showed that in that period the procyclicality of consumption was much reduced showing the output correlation .51, and that of investment still more reduced, with the output correlation .16, while the procyclicality of employment remained high having the output correlation .78. The dynamics constructed in this book reproduces also this deviation and part of the similar deviations in Britain rather well, as will be shown in Chapter 9.

1.2 This Theory Fits Well the Facts But Does It Fit the Current Dogmas?

The central dogma of contemporary macroeconomics demands a reduction of macroeconomics to the free microeconomic exchange of paid goods and paid services between firms and households. Its current rigidity is quite recently born. Still in 1985 the leading textbook of mainstream economics could state: "First, contrary to what its early critics believed, the Soviet economy has grown rapidly, has expanded its influence, and has won many allies ... From the point of view of economics, perhaps the most significant lesson is that a command economy can function."[10]

The rigidity of the reductionist dogma in the 1990s suggests that the crash of the Soviet Union helped to stress the importance of the free exchange of goods and services in a healthy economy. But the totalitarian nature of the whole Soviet system should not be forgotten

[9]Correia,I., J.Neves and S.Rebelo, Business cycles from 1850 to 1950, European Economic Review 36, 1992, p.459-467, Table 2.

[10]Samuelson,Paul and William Nordhaus,Economics,MacGraw-Hill 1985, p.771.

either: government controlled human life from birth to death, allowing people to have indeed no real time for themselves. This destroyed the necessary individual freedom that seems to be a condition sine qua non of genuine endogeneous growth.

The remarkable economic success of Western countries recorded by Angus Maddison[11] will be in this book interpreted in terms of the parallel development of Western natural science and endogeneous economic growth in a way which stresses the importance of individual freedom for both of these pursuits. Thus *command economies are excluded in theory as well as in praxis.* The mathematical formulation of this idea, given in Chapter 3, proves to define a minimum macroeconomic dynamics, necessary to explain the facts 1-17. The ideas underlying this dynamics will be discussed in more detail in Section 1.3 below[12].

Am I mistaken when I see also signs of a second dogma in the process of emanating from the demand that macroeconomic theory should be given the form of stochastic processes? Such a demand would deliberately reduce the level of causal explanation in economics, as correctly remarked by Richard Day[13]). The first synod declaring the new dogma seems to have been a recent book (mentioned in footnote 5). But while the stochastic approach – method (a) – is a possible logical choice in the theory of growth cycles, it does not explain the basic quantitative facts 10-15 concerning the ordinary business cycles or the facts 16 and 17 on anomalous growth and growth cycles.

This is one of the reasons why the method (b), leading to classical canonical formalism with an invariance group, and a differential rather than difference calculus, is here applied instead of the stochastic method (a). The shocks are superposed on causal cycles but prove to be of minor importance.

Both of those methodical orientations, which were above called 'dogmas', are now accepted and applied by the majority of research economists. Therefore a more detailled verbal introduction to the different position taken in this book will be given in Section 1.3.

---

[11] Angus Maddison, Dynamic Forces in Capitalist Development, OUP 1991.

[12] For a more extensive discussion see A.Aulin, The Origins of Economic Growth, Springer-Verlag 1997, Part I.

[13] Richard H.Day, Review article, Structural Change and Economic Dynamics 3, 1992, p. 177-182.

1.3 Why This Dynamics Explains the Facts While the Current Dogmas Fail So Badly?

Reason No.1: Generalized Human Capital

The success of the theory constructed in Chapter 3 in the explanation of facts 1-17 is spectacular as compared with the failure of the usual methods based on microeconomics and stochastic processes. The best way to understand this is to study the parallel history of exact science and economic growth, starting from the concept of human capital.

In economics human capital is defined as a factor of production. It stands for the economically profitable average knowledge and skills of labour force in society. But it is next to impossible to tell which part, for instance, of scientific knowledge will be economically profitable in future. Therefore we must start out by considering a generalized concept of human capital. This will give a part of the answer to the above question.

The beginning of the period of a more or less permanent endogeneous economic growth in advanced countries roughly coincides, historically, with the birth of exact natural science in its modern sense. Both the endogeneous economic growth and a large-scale accumulation of exact knowledge in natural science started within Western civilization. Both of them had their roots in the Middle Ages and, as far as science is concerned, even in the antiquity. However a larger growth and accumulation did not begin before the modern times.

Scientific intellect of course had produced interesting innovations even before the beginning of modern exact science in 16th and 17th centuries. Mathematics, particularly in the form of geometry and elementary number theory, was born much earlier and had flourished in the Greek and Hellenic cultures of the antiquity.The Arabic culture produced the elements of algebra and the Arabic numerals. And, to quote Angus Maddison[14], "medieval innovations had included windmills, horseshoes, horse harness, heavy ploughs, the haystack, the scythe, marling, fertilization, and the three-field rotation system. These inno-

[14] Angus Maddison, Dynamic Forces in Capitalist Development, Oxford University Press, Oxford 1991, p.14.

vations spread rather gradually, but they were certainly helping to increase agricultural output in northern Europa." This is clearly reflected in the growth of yield ratios in Western Europe.[15]

However, "innovation was much slower in this epoch than it is now because the main locus of production was in agriculture, where innovation was too risky for most of the participants and often inhibited by tenure institutions ... in urban handicrafts, guild restrictions also limited possibilities for change. Entry to skilled occupations was carefully controlled and technical knowledge was regarded as a mystery not to be shared with those outside the recognized fraternity." (Maddison, ibid.). Therefore the discovery of printing around 1500 was a major step in the diffusion of scientific and technological knowledge.

In the 18th and 19th centuries maturated exact natural sciences, in particular physics and chemistry. This early development of exact science had already a great influence on technology and economic practice. Steam machine, built on the foundation of classical mechanics and thermodynamics, mechanized the pumps in mines since the 18th century. In 1804 it moved the first locomotive and in 1807 chugged the first boat using the steam machine up the Hudson river. Electromagnetism took a great leap forward in the 19th century, and produced as its applications the electric motor, electric light and electric telegraph. Modern quantum mechanics has had its many applications in electronics, computers and telecommunications. Modern chemistry, biochemistry and even microbiology are all of them now, in the last analysis, applications of quantum electrodynamics.

By comparing the realized economic growth, say between 1400 and 1989, or between 1820 to 1989, in the countries where the breakthrough of exact science took place, with the realized growth in other countries, we can see the difference. According to Maddison[16] the GDP per capita in Western Europe and its offshoots (Australia, Canada and the USA) was in 1989 33.5 times its GDP per capita in 1400, while in China it had multiplied only by a factor of 4.7. Thus an average citizen in an advanced country was 33 times as wealthy in 1989 as an average citizen was in 1400. And if we take the 16 economically most advanced

[15]B.H. Slicher van Bath, Yield ratios 810-1820, Afdeling Agrarische Geschiedenis, Bijdragen No.10, Wageningen 1963.

[16]A.Maddison, ibid., p.10.

countries, from Western Europe and its offshoots plus Japan, we find on the average a 14-fold GDP per capita in 1989 when compared with what it was in 1820. The progress of exact science in these countries from 1400 to 1989 or from 1820 to 1989 is not measurable but we can guess: it was enormous when compared with the countries which did not partake in this progress.

The accumulation is equally characteristic of the progress of scientific knowledge in exact sciences as it is of economic growth. The prototype and also the core of all accumulating exact scientific knowledge is the fundamental theoretical knowledge in mathematical sciences. Here the accumulation of theoretical knowledge concerning the causal explanation of observed phenomena is seminal. The accumulation of this kind of fundamental theoretical knowledge is striking in the basic natural science, physics.

Thus, for instance, classical physical theory was generalized by Einstein's theory of relativity in a way that retained the old theory valid in a special case, viz. for low velocities. The magnitude of theoretical knowledge was in this process accumulated: the new theory only added something to the earlier bulk of theoretical knowledge, not refused it. We meet a similar accumulation of existing knowledge in another important case in point, viz. in quantum theory. This too was a generalization of classical ("Newtonian") physical theory but in another direction, from macrophysics to microphysics. Classical theory remained valid in macrophysics. Thus again the fundamental physical theory was accumulated: nothing had been taken out of the earlier knowledge, only new fundamental elements were added to the bulk of causal knowledge now understood in a probabilistic sense.

The development of exact scientific knowledge indeed means accumulation of this knowledge. This accumulation may even be the primary one, and the accumulation of output in endogeneous economic growth may be secondary to it. After all the latter is in the long run mainly based on the exploitation of exact sciences in technology.

The more advanced is a society, the closer link in history there seems to be between the growth of knowledge in science and the growth of economy. Therefore there are reasons to consider a

> *generalization of human capital* from the immediately economically profitable one to a concept referring to all accu-

> mulated and exact scientific knowledge, and to the skills based on it, all of which may potentially produce economic growth, if not immediately then later.

Evidently, human capital in this sense is increased only by creation of new knowledge. The economists have mostly identified human capital with education. But education implies only transmission of knowledge from one person to another, which of course helps economic growth but does not create new knowledge.

Reason No.2: Generalized Value Function

Scientific innovations and the technological innovations based on them of course do not come from a vacuum. They become possible only in the course of an active pursuit of scientific or technological progress as a value of its own. Genuine innovations are not made to order at any desired point of time but they come when they come, possibly during hours devoted to some kind of work. But they may be born also during the leisure time. In fact we can expect that most of the scientific and technological innovations start from some ideas, which originally come to a well trained mind during the free wanderings of thought characteristic of the leisure time. Thus we obviously cannot exclude leisure from a scientific formalism discussing the economically relevant use of human time.

Secondy, the pursuit of scientific or technological innovations, considered as a *nonmaterial value*, has no finite upper limit. Indeed we can increase new scientific knowledge and cognitive innovations without any limit, and all this new knowledge may potentially create economic growth, if not immediately then later. In this respect the growth of nonmaterial values involved in science and cognitive innovations radically differes from material consumption, whose growth typically improves the wellbeing of a person less and less until no further consumption improves it.

It follows that the utility function standing for nonmaterial values involved in the pursuit of innovations has to be *unbounded*.

Material and nonmaterial values certainly often mix in human mind, but the 'purely' nonmaterial part of any such value can be represented in economic utility only by a leisure term having an unbounded utility function. Such a leisure term itself can be interpreted to stand for the unbounded value of individual freedom. But it also makes possible the representation of the economic effects of other 'purely' nonmaterial values, thus defined as inexhaustible values pursued during the leisure time.

All this demands an extension of the value function to include also nonmaterial values. Therefore the usual representation of technological innovations merely by factors in the production function is insufficient – how ever inventive it is (as in the case of Evans,Honkapohja and Romer,1996). In Chapter 3 the following idea will be given a quantitative, in economic terms measurable form, to be empirically verified in Part 2 of this book:

> *Thesis.* Economic development must be seen in a wider context of human development in which the accumulation of exact scientific knowledge plays a central role. But innovations in exact science involve, besides the pursuit of material values, also the pursuit of nonmaterial values such as truth, intellectual beauty etc. Hence the macroeconomic value function must be generalized in a way that makes it possible to represent also the economic effects of nonmaterial values.

It proves that

1) the extension to involve nonmaterial values, included in the dynamics of this book and leading to the existence of an invariance group of the fundamental equations, is necessary for the explanation of the basic facts 1-17, while

2) such an extension is impossible in the approaches based on microeconomics superimposed by stochastic processes, which is why such approaches and thus the current dogmas do not and cannot explain the basic macroeconomic facts.

Monetary movements are not discussed in this book.

Part 1

THE FUNDAMENTAL LAWS OF SCIENCE-BASED ECONOMIC GROWTH

2 Canonical Formalism: Basics

The canonical formalism will be here revisited emphasizing the parametric conditions of the existence of solutions in economic applications.

2.1 Elementary Mathematics

Hamiltonian Formalism

Assuming all the necessary properties of existence and continuity we write the Hamiltonian equations as usually,

$$(2.1)\qquad \dot{q}_\nu = H_{p_\nu}, \quad \dot{p}_\nu = -H_{q_\nu}, \qquad (\nu = 1, 2, ..., r)$$

for the position co-ordinates q_ν and the momentum components p_ν. The Hamiltonian function $H(q_\nu, p_\nu)$ does not depend explicitly on time. It follows that it is a constant of motion:

$$(2.2)\qquad \dot{H} = \sum_{\nu=1}^{r} \left(H_{q_\nu} \dot{q}_\nu + H_{p_\nu} \dot{p}_\nu \right) = 0.$$

A formal solution of (1) can be written, obviously, as

$$\begin{aligned} q_\nu &= e^{tD} q_\nu(0), \quad p_\nu = e^{tD} p_\nu(0), \qquad (\nu = 1, 2, ..., r), \\ D &= \sum_{\nu=1}^{r} \left(H_{p_\nu} \partial_{q_\nu} - H_{q_\nu} \partial_{p_\nu} \right). \end{aligned}$$

The differential operator D gives the time derivative of any function $z(q_\nu, p_\nu)$: $\dot{z} = Dz$. The total state of the Hamiltonian system is defined by the totality of the components q_ν and p_ν.

A system of n mass points, moving in the 3-dimensional physical (Euclidean) space according to the laws of Newtonian mechanics, is the prototype of a Hamiltonian system. Each of the mass points, consisting of matter concentrated at a point,has three position co-ordinates and three momentum components, thus $r = 3n$. The momentum components of each mass point are defined as the product of its mass and its respective component of velocity, and can be accordingly written as

$$p_\nu = m_\nu \dot{q}_\nu, \qquad (\nu = 1, 2, ..., r) \tag{2.3}$$

where the same mass m_ν, of course, appears three times. Then the equations of motion are given by (1), with the Hamiltonian function

$$H = T + V, \quad T = \sum_{\nu=1}^{r} \frac{1}{2m_\nu} p_\nu^2, \quad V = V(q_\nu). \tag{2.4}$$

Here T, V and H are the kinetic, potential and total energy, respectively, of the system.

Because of (3) the laws of motion of a system of mass points can also be expressed in terms of the q_ν and $\dot{q}_\nu$ instead of the q_ν and p_ν. Instead of the Hamiltonian equations (1) we then have the *Euler equations of motion*,

$$\frac{d}{dt} L_{\dot{q}_\nu} = L_{q_\nu}, \qquad (\nu = 1, 2, ..., r). \tag{2.5}$$

The Lagrangian function L gives the difference between the kinetic and potential energies, i.e.

$$L(q_\nu, \dot{q}_\nu) = \frac{1}{2} \sum_{\nu=1}^{r} m_\nu \dot{q}_\nu^2 - V(q_\nu). \tag{2.6}$$

The formulae (1) and (5) of course are mutually equivalent expressions of the equations of motion of the mass point system. The transformations $H \to L$ and $L \to H$, or from the variables (q_ν, p_ν) to the variables $(q_\nu, \dot{q}_\nu)$ and vice versa, can be given the following symmetric form:

$$L(q_\nu, \dot{q}_\nu) + H(q_\nu, p_\nu) = \sum_{\nu=1}^{r} \dot{q}_\nu p_\nu, \tag{2.7}$$

$$L_{\dot{q}_\nu} = p_\nu, \quad H_{p_\nu} = \dot{q}_\nu. \tag{2.8}$$

The latter equations obviously follow from (1) and (5), while (7) has the double kinetic energy on both sides. From (7) we immediately obtain the further relation

$$L_{q_\nu} + H_{q_\nu} = 0.$$

The Origins of Canonical Formalism

William Hamilton also constructed, using the variational calculus developed by the Swiss mathematician Jean Bernoulli, a more general dynamics called 'canonical formalism'.

We now consider an $r+1$-dimensional space of the variables u_ν and the time t. Let the u_ν be functions of time, continuous and continuously differentiable. Let A and B be two points of this space, defined by $A = (u_\nu(0), 0)$ and $B = (u_\nu(t_1), t_1)$, $t_1 > 0$. We let B move freely on the surface $T(u_\nu(t_1), t_1) = 0$.

For any function $L(u_\nu, \dot{u}_\nu, t)$, continuous and at least twice differentiable with respect to its variables, and satisfying the condition

$$|L_{\dot{u}_\nu \dot{u}_\mu}| \neq 0,$$

where the expression $|\cdot|$ means determinant, the geodetic distance from A to B is defined as the smallest value of the integral

$$J(u_\nu(t_1), t_1) = \int_0^{t_1} L(u_\nu, \dot{u}_\nu, t)\, dt. \tag{2.9}$$

A necessary condition of an extremal value of J is obtained by variation of the path $u_\nu(t)$ from A to B and requiring that $\delta J = \int_0^{t_1} \delta L dt = 0$. But

$$\delta L = \sum_{\nu=1}^{r} \left(L u_\nu \delta u_\nu + L_{\dot{u}_\nu} \delta \dot{u}_\nu \right),$$

where, by partial differentiation, we have

$$L_{\dot{u}_\nu} \delta \dot{u}_\nu = \frac{d}{dt} \left(L_{\dot{u}_\nu} \delta u_\nu \right) - \frac{d}{dt} L_{\dot{u}_\nu} \delta u_\nu.$$

This gives, since $\delta u_\nu(0) = 0$:

$$\delta J = \sum_{\nu=1}^{r} \int_0^{t_1} \left(L u_\nu - \frac{d}{dt} L_{\dot{u}_\nu} \right) \delta u_\nu \, dt + \sum_{\nu=1}^{r} L_{\dot{u}_\nu}(t_1) \delta u_\nu(t_1) = 0.$$

Since this must hold good for any points on the path and since $\delta u_\nu(t_1)$ is different from zero at least for some value of ν, the first term gives the Euler equations now written for the function L with an explicit time-dependence,

$$\frac{d}{dt}L_{\dot{u}_\nu}(u_\nu, \dot{u}_\nu, t) = L_{u_\nu}(u_\nu, \dot{u}_\nu, t), \qquad (\nu = 1, 2, ..., r)$$

while the second term defines what is called the natural boundary conditions:

(2.10) $$L_{\dot{u}_\nu}(t_1) = 0. \qquad (\nu = 1, 2, ..., r)$$

If we drop the explicit time-dependence of the action function L, and if we put $u_\nu = q_\nu$, we can identify L with the Lagrange function of a system of mass points. Thus interpreted the proof given in this section has shown that the motion of particles in such a system takes place along the "least action" path.

In imitation of the equations (7) and (8) holding good for a mass point system, we can define for any given action function $L(u_\nu, \dot{u}_\nu, t)$ now the canonical momentum v_ν, associated with the function u_ν, by writing

(2.11) $$v_\nu \overset{def}{=} L_{\dot{u}_\nu}, \qquad (\nu = 1, 2, ..., r)$$

and the Legendre function G by the formula

(2.12) $$G(u_\nu, v_\nu, t) + L(u_\nu, \dot{u}_\nu, t) \overset{def}{=} \sum_{\nu=1}^{r} \dot{u}_\nu v_\nu.$$

In the imitation, the rest of the transformation (7)-(8) gives

(2.13) $$\dot{u}_\nu = G_{v_\nu}. \qquad (\nu = 1, 2, ..., r)$$

By derivation, (12) immediately gives

(2.14) $$G_{u_\nu} + L_{u_\nu} = 0.$$

It follows that the Euler equations of the action principle are equivalent to the equations

(2.15) $$\dot{u}_\nu = G_{v_\nu}, \quad \dot{v}_\nu = -G_{u_\nu}, \qquad (\nu = 1, 2, ..., r)$$

called the canonical equations of motion.

It is to be emphasized that in these equations both functions G and L in the general case depend explicitly on time. Thus the canonical dynamical system is in the general case not a Hamiltonian system and it has necessarily no constant of motion.[1]

However, just like in a Hamiltonian system to every Lagrangian function there corresponds a Hamiltonian function and vice versa, in a canonical system *to every variational integrand or the action function* $L(u_\nu, \dot{u}_\nu, t)$ *there corresponds a Legendre function* $G(u_\nu, v_\nu, t)$ *and vice versa.*

For a small time-displacement δt_1 in (9) we get, in view of (12):

$$\delta_{t_1} J = L(t_1)\delta t_1 = \left[\sum_{\nu=1}^{r} \dot{u}_\nu(t_1) v_\nu(t_1) - G(t_1)\right] \delta t_1.$$

Similarly, for a small displacement of the path $u_\nu(t)$ we first get by applying (12):

$$\delta_U J = \int_0^{t_1} \sum_{\nu=1}^{r} \left[v_\nu \delta\dot{u}_\nu + \dot{u}_\nu \delta v_\nu - G_{U_\nu} \delta u_\nu - G_{V_\nu} \delta v_\nu\right] dt.$$

Here the canonical equations (15) can be applied to give

$$\delta_U J = \sum_{\nu=1}^{r} \int_0^{t_1} \frac{d}{dt}(v_\nu \delta u_\nu)\, dt = \sum_{\nu=1}^{r} v_\nu(t_1) \delta u_\nu(t_1). \tag{2.16}$$

(Note that $\delta u_\nu = 0$ for $t = 0$.)

Written for a variable time t instead of the fixed time t_1 the variational formulae give:

$$\frac{dJ}{dt} = \sum_{\nu=1}^{r} J_{U_\nu} \dot{u}_\nu - G, \quad \text{thus } \partial_t J = -G.$$

Hence we get the Hamilton-Jacobi equation

$$\partial_t J + G(J_{U_\nu}, u_\nu, t) = 0. \tag{2.17}$$

[1] In dynamic economics the dynamical systems usually are canonical systems, not Hamiltonian ones. Therefore no constant of motion is necessary.This has been sometimes forgotten, for instance when P.Mirowski (1990,p.302) regrets that the economists "have never made up their minds about what precisely it is that should be conserved in their theoretical system".

It defines the geodetic distance from A to B as a function of the moving end point $B = (u_\nu(t), t)$.

On the other hand, the surface of equivalence of the geodetic distance J from the fixed initial point A to the point B is given by

$$J(u_\nu(t_1), t_1) = \text{Constant},$$

or, written for the variation δJ:

$$\delta J(t_1) = \sum_{\nu=1}^{r} (J_{u_\nu})_{t_1}\, \delta u_\nu(t_1) + (\partial_t J)_{t_1}\, \delta t_1 = 0. \tag{2.18}$$

In order for the extremal path $u_\nu(t)$ to be transverse with respect to a given boundary surface $T(u_\nu(t_1), t_1) = 0$ through the point B, the equation of the surface T, in a differential form

$$\delta T(t_1) = \sum_{\nu=1}^{r} (T_{u_\nu})_{t_1}\, \delta u_\nu(t_1) + (\partial_t T)_{t_1}\, \delta t_1 = 0,$$

must be equivalent to (18) in a neighbourhood of B, i.e. there must be a proportionality of the coefficients:

$$(\partial_t J)_{t_1} : (J_{u_\nu})_{t_1} = (\partial_t T)_{t_1} : (T_{u_\nu})_{t_1} . \qquad (\nu = 1, 2, ..., r)$$

These relations are called the conditions of transversality of the extremal path $u_\nu(t)$ with respect to the given surface $T = 0$. In view of (16) and (17) they can be written in the form

$$\left(\frac{-G}{v_\nu}\right)_{t_1} = \left(\frac{\partial_t T}{T_{u_\nu}}\right)_{t_1} . \qquad (\nu = 1, 2, ..., r) \tag{2.19}$$

They too are, of course, only necessary but not sufficient conditions of the attainment of the minimal distance from A to B.

A further necessary but not sufficient condition of the minimal distance is the Legendre condition

$$\sum_{\nu,\mu=1}^{r} \dot{u}_\nu | L_{\dot{u}_\nu \dot{u}_\mu} | \dot{u}_\mu \geq 0 \quad \text{for} \quad 0 \leq t < t_1. \tag{2.20}$$

For the derivation of this condition a consultation of the well-known advanced textbook of Courant-Hilbert, Methods of Mathematical Physics

I-II (in its English translation first printed by Interscience Publishers, New York, in 1953) is suggested.

If instead of a minimum J a maximum of J is wanted, we simply have to replace the variational integrand L by $-L$ in all the above formulae. The Legendre transformation (11)-(13) then reads:

$$v_\nu = -L_{\dot{u}_\nu}, \quad \dot{u}_\nu = G_{v_\nu}, \qquad (\nu = 1, 2, ..., r) \tag{2.21}$$

$$G(u_\nu, v_\nu, t) = L(u_\nu, \dot{u}_\nu, t) + \sum_{\nu=1}^{r} \dot{u}_\nu v_\nu. \tag{2.22}$$

Thus we have now

$$G_{u_\nu} = L_{u_\nu}, \qquad (\nu = 1, 2, ..., r) \tag{2.23}$$

and the canonical equations of motion retain their form (15).

The transversality conditions (19) and the natural boundary conditions (10) are unchanged, while the Legendre condition (20) obviously is changed to

$$\sum_{\nu,\mu=1}^{r} \dot{u}_\nu |L_{\dot{u}_{nu}\dot{u}_\mu}| \dot{u}_\mu \leq 0 \quad \text{for} \quad 0 \leq t < t_1. \tag{2.24}$$

It is in the form of the "principle of the largest action" that the canonical formalism is applied in economics.

2.2 The Use of Canonical Formalism in Economics

Maximization of the Value Function

Suppose we have to maximize the value function

$$J(u_\nu(t_1), t_1) = \int_0^{t_1} e^{-\rho t} U(u_\nu, \dot{u}_\nu, t) dt, \quad \rho > 0,$$

where U is the current-time utility function depending on a number of factors of production u_ν, their time derivatives and the time t, ρ being the discount rate.

The problem is one of the action principle applied to discounted utility

$$L(u_\nu, \dot{u}_\nu, t) = e^{-\rho t} U(u_\nu, \dot{u}_\nu, t). \tag{2.25}$$

For L the Legendre transformation (21)-(22) is valid: the canonical momentum v_ν is now defined by

$$v_\nu = -L_{\dot{u}_\nu} = -e^{-\rho t} U_{\dot{u}_\nu}, \qquad (\nu = 1, 2, ..., r) \tag{2.26}$$

the Legendre function G by

$$G(u_\nu, v_\nu, t) = e^{-\rho t} U(u_\nu, \dot{u}_\nu, t) + \sum_{\nu=1}^{r} \dot{u}_\nu v_\nu, \tag{2.27}$$

and the canonical equations are given by

$$\dot{u}_\nu = G_{v_\nu}, \quad \dot{v}_\nu = -G_{u_\nu} = -L_{u_\nu}. \tag{2.28}$$

Let us introduce, with Kurz (1968), the variables

$$\pi_\nu \stackrel{def}{=} -U_{\dot{u}_\nu} = e^{\rho t} v_\nu, \qquad (\nu = 1, 2, ..., r) \tag{2.29}$$

$$G^* \stackrel{def}{=} e^{\rho t} G(u_\nu, v_\nu, t). \tag{2.30}$$

In terms of these variables the canonical equations (15) become

$$\dot{u}_\nu = e^{-\rho t} G^*_{v_\nu} = G^*_{\pi_\nu}, \qquad (\nu = 1, 2, ..., r) \tag{2.31}$$

$$\dot{\pi}_\nu = \rho \pi_\nu - G^*_{u_\nu}. \qquad (\nu = 1, 2, ..., r) \tag{2.32}$$

These equations are in economics often called the "modified Hamiltonian equations", which is misleading in two ways: first, it misleads one to believing that the function G^* is a Hamiltonian function, which it is not; secondly, it wrongly suggests that the system considered is a Hamiltonian system, which it is not. The system is a non-Hamiltonian canonical system, and the function G^* of course belongs to the category of Legendre functions. Let us call it the current-time Legendre function.

The transversality conditions retain their form (19), only the definitions of the variable v_ν and the function G have changed to the forms (26) and (27), respectively. When expressed in terms of the Kurz momentum π_ν the conditions are:

$$\left(\frac{-G}{e^{-\rho t}\pi_\nu}\right)_{t_1} = \left(\frac{\partial_t T}{T u_\nu}\right)_{t_1}. \qquad (\nu = 1, 2, ..., r) \tag{2.33}$$

The Legendre condition is that of the "principle of the largest action", i.e. (24). The natural boundary conditions (10) too are unchanged but can be now written in the form

$$\left(e^{-\rho t} U_{\dot{u}_\nu}\right)_{t_1} = 0. \qquad (\nu = 1, 2, ..., r) \tag{2.34}$$

Mostly in applications to growth theory $t_1 = \infty$.

Example: The Solow Model

We have a single factor of production, the physical capital K, together with a given labour force $N = N(0)e^{nt}, n > 0$, to produce the output Y and the consumption per capita c by means of the following production function and growth equation:

$$Y = AK^\beta N^{1-\beta}, \quad \dot{K} = sY = Y - cN, \quad \dot{A}/A = g > 0, \tag{2.35}$$

$$c = \frac{1}{N}\left(AK^\beta N^{1-\beta} - \dot{K}\right) = c(K, \dot{K}, t), \quad 0 < \beta < 1. \tag{2.36}$$

We maximize

$$J(K, t_1) = \int_0^{t_1} e^{-\rho t} N V(c(K, \dot{K}, t))dt = N(0)\int_0^{t_1} e^{(n-\rho)t} V(c)dt, \quad t_1 = \infty, \tag{2.37}$$

where the current-time utility per capita V must be determined so that the Legendre condition and the natural boundary condition are satisfied.

The canonical Kurz momentum p associated with the variable K, according to (26),(29) and (35), is

$$p = -N(V_c)\, c_{\dot{K}} = V_c, \tag{2.38}$$

By imposing on $V(c)$ the further assumption that it is a bijection, we have

$$c = V_c^{-1}(p), \tag{2.39}$$

which expresses the consumption per capita in terms of the Kurz momentum p.

The current-time Legendre function G^* is then, according to (27), (30),(37) and (38), given by

$$G^*(K,p,t) = NV(c(p)) + p\dot{K}(K,p,t), \tag{2.40}$$

where, in view of (11),

$$\dot{K}(K,p,t) = Y(K,t) - c(p)N(t).$$

It follows that the canonical equations of motion, (31) and (32), now have the form

$$\begin{aligned} (2.41)\; \dot{K} &= Y(K,t) - c(p)N(t), \qquad \text{(growth equation of capital)} \\ (2.42)\; \dot{p} &= \rho p - \beta p Y/K. \qquad \text{(Euler equation)} \end{aligned}$$

The natural boundary condition and the Legendre condition now become

$$\lim_{t\to\infty} e^{-\rho t} V_c = 0, \tag{2.43}$$

$$V_{cc} < 0 \quad \text{for finite times } t \geq 0, \tag{2.44}$$

respectively.

The "balanced-growth" substitutions

$$Y \to Y^* = Y^*(0)e^{\lambda t}, \quad K \to K^* = K^*(0)e^{\lambda t},$$

where

$$\lambda = s^* b^* = n + \frac{g}{1-\beta}, \tag{2.45}$$

s^* and b^* being positive constants, viz. the balanced-growth net savings rate and the balanced-growth net output/capital ratio, respectively, obviously satisfy the growth equation in (35), equivalent to (41). This gives

$$c \to c^* = b^*(1-s^*)[K^*(0)/N(0)]e^{(\lambda - n)t} \to \infty \quad \text{with } t \to \infty$$

since, because of (45),

$$\lambda > n. \tag{2.46}$$

It follows that the choice

$$V(c) = \frac{1}{1-\sigma}\left(c^{1-\sigma} - 1\right) \tag{2.47}$$

for the function V satisfies (43) and (44) for any positive σ. (For $\sigma = 1$ it gives $\log c$.) Here the positive constant σ can be interpreted as the coefficient of risk aversion (Lucas,1988). With this choice for the function $V(c)$ we have, in view of (38),

$$p = c^{-\sigma}. \tag{2.48}$$

To satisfy also the Euler equation (42) on the balanced-growth path we must have, along with (48), the further parameter condition

$$\rho + \sigma(\lambda - n) = \beta b^*. \tag{2.49}$$

When solved for σ this gives

$$\sigma = \frac{\beta b^* - \rho}{s^* b^* - n}. \tag{2.50}$$

Thus the risk aversion coefficient is not an independent parameter but is determined by β, b^*, s^*, ρ and n. In (50) both the nominator and denominator are positive, because of (45) and (46).

Let us study under which conditions the balanced-growth solution satisfies the transversality condition. It is natural to choose the surfaces of equivalence of output, $Y(K,t) = \text{Constant}$, as the boundary surfaces

$T = 0$. This gives, in view of (33) and (35), the following transversality condition:

$$\left(\frac{-G}{e^{-\rho t}p}\right)_{t_1} = \left[\frac{g + n(1-\beta)}{\beta}\right] K(t_1), \quad \text{for } t_1 = \infty. \tag{2.51}$$

In this form the condition is not valid in the Solow model, since for $t_1 = \infty$ the left-hand side gives negative infinity, while the right-hand side gives the positive one. By rewriting it as

$$\left[\frac{g + n(1-\beta)}{\beta}\right] e^{-\rho t} p(t_1) K(t_1) = -G(t_1), \quad \text{for } t_1 = \infty, \tag{2.52}$$

this problem is avoided.

First we shall study under what conditions, to be imposed on parameters, the transversality condition (52) of the Solow model can be given in its usual "textbook form"

$$e^{-\rho t} p(t) K(t) \to 0 \quad \text{with } t \to \infty. \tag{2.53}$$

By applying (30),(40) and (47) we get

$$G = N(0) e^{(n-\rho)t} \left(\frac{c^{1-\sigma} - 1}{1-\sigma}\right) + e^{-\rho t} p\dot{K}. \tag{2.54}$$

There are two alternative cases:

Case (i). On the condition that

$$\sigma > 1 \tag{2.55}$$

we have

$$\frac{c^{1-\sigma} - 1}{1-\sigma} \to \text{Constant} \quad \text{with } t \to \infty.$$

On this condition the first term in (54) approaches zero with time, provided that the further condition

$$\rho > n \tag{2.56}$$

is also satisfied. From (55) and (56) it follows, in view of (50), that also the following parameter relation holds good:

(2.57) $$s^* < \beta.$$

But this is just the condition under which the second term in (54) vanishes in the limit $t \to \infty$. Thus (55) and (56) express the necessary and sufficient conditions under with the transversality condition of the Solow growth model assumes the "textbook form" (53).

Case (ii). If (55) is not valid, the maximal utility per capital function will be obtained for $\sigma = 0$, in which case

$$\frac{c^{1-\sigma} - 1}{1-\sigma} \to c - 1.$$

This approaches infinity on the balanced-growth path like the exponential function $\exp(\lambda - n)t$. It follows that the first term in (54) will approach zero with increasing time on the condition that we have now, instead of (55),

(2.58) $$\rho > \lambda.$$

This is a very strong condition, stronger than (56) and, in view of empirical evidence, also stronger than (55). In order that the "textbook form" would express the transversality condition in the Case *(ii)*, both parameter relations (57) and (58) must hold good.

We have so far studied under which conditions the transversality relation of the Solow model can be given its "textbook form" (53). But the same conditions already certify the validity of this relation on the balanced-growth path. This is because the parameter condition (57) is also sufficient to make the expression $e^{-\rho t}p^*K^*$ to approach to zero asymptotically. (This may evoke the question: why to write at all the "textbook version" of the transversality condition, since its validity is already guaranteed when writing it down? A potential objection to this question could be that the conditions of convergence might be different in the cases, where the initial state $K(0)$ is not on the balanced-growth path. But this is not true. On the contrary, for any initial capital $K(0) > 0$ the general solution of the Solow model converges to the balanced-growth path asymptotically.)

The importance of the parameter relations (46),(49),(50),(55) and (56) has been emphasized above, since the same relations will appear also in the Lucas growth model and in its generalization later on. A further relation of considerable interest is obtained when solving the balanced-growth Euler equation (49) for the parameter b^*:

$$b^* = \frac{n - \rho/\sigma}{s^* - \beta/\sigma}. \tag{2.59}$$

We shall have reason to return also to this formula later.

The Arrow-Kurz Generalization

From (40),(41) and (48) we get:

$$G^*_c = N(V_c - p) = 0 \quad \text{for} \quad p = V_c,$$
$$G^*_{cc} = NV_{cc} < 0.$$

Thus the Legendre function G^* attains a maximum with respect to c at the point, where V_c is equal to the Kurz momentum p.

If we forget now the origin of this Legendre function, and *define* a surrogate function H^* by writing

$$H^*(K,p,t;c) \overset{def}{=} NV(c) + p\dot{K}(K,t,c), \quad \dot{K} = Y(K,t) - cN, \tag{2.60}$$

we can perform the maximization of the accumulated utility

$$J = \int_0^{t_1} e^{-\rho t} NV(c)dt$$

in the following simple way.

First choose $c = \hat{c}$, where $\hat{c}$ solves $H^*_c = 0$ and $H^*_{cc} < 0$, to get a function H^o of the variables K, p and t:

$$H^o(K,p,t) = NV(\hat{c}) + p\dot{K}(K,t,\hat{c}). \tag{2.61}$$

Then, obviously,

$$H^o_p = \dot{K}. \tag{2.62}$$

Consider then the Euler equation

$$\frac{d}{dt}L_{\dot{K}} = L_K,$$

which is a necessary condition of the maximization of the accumulated utility J, provided that

$$L(K, \dot{K}, t) = e^{-\rho t} N V(c). \tag{2.63}$$

Hence we get successively:

$$\begin{aligned} \frac{d}{dt} L_{\dot{K}} &= \frac{d}{dt} \frac{\partial}{\partial \dot{K}} e^{-\rho t} N V(c) \\ &= \rho e^{-\rho t} V_c - e^{-\rho t} \frac{d}{dt} V_c, \\ L_K &= e^{-\rho t} N V_c c_K = e^{-\rho t} V_c Y_K. \end{aligned}$$

For $c = \hat{c}$ this gives

$$\dot{p} = \rho p - H^o_K. \tag{2.64}$$

Thus we can maximize the accumulated utility also by

1) maximizing H^o with respect to the control parameter c and

2) writing down the "modified Hamiltonian equations" for the surrogate function H^o, in this connection often called the "current-time Hamiltonian".

Arrow and Kurz (1970) showed that this shorthand method can be as well used in a many-factor and many-parameter case, to maximize

$$J = \int_0^{t_1} e^{-\rho t} U(u_\nu, \dot{u}_\nu, t; a_\mu) dt, \tag{2.65}$$

subject to the growth equations

$$\dot{u}_\nu = f^{(\nu)}(u, t; a_\mu), \qquad (\nu = 1, 2, ..., r; \mu = 1, 2, ..., m), \tag{2.66}$$

where $u = (u_1, u_2, ..., u_r)$. Thus one must first write down the equations $\partial H^o / \partial a_\mu = 0$, and then add to the list of equations the "modified Hamiltonian equations" for each pair of the canonical variables (u_ν, π_ν). The three kinds of necessary conditions, viz. the natural boundary conditions, the Legendre condition and the transversality conditions must of course be finally added to the list. This generalized method will be applied in this book.

Example: The Lucasian Mechanics of Economic Growth

The equilibrium model of the Lucas (1988) mechanics of economic growth has a particular two-phased structure. First there is a situation, in which the households and firms react to what is expected to be the current level of the economically profitable technological knowledge and skills h_a in society. In the theory this level is first exogeneously given, just as the exogeneous factor of technological progress in the Solow model. The fundamental equations of the theory are constructed in this *phase 1*, which can be called the "reaction of the market to the expected level of knowledge and skills in society".

The market clearing then creates a second situation, the *phase 2*, in which the expected level of human capital and the endogeneously produced average level h of human capital coincide. The solution of the fundamental equations has to take place in this situation.

We maximize

$$(2.68) \quad J = \int_0^\infty L dt = \int_0^\infty e^{-\rho t} N \left(\frac{c^{1-\sigma} - 1}{1-\sigma} \right) dt, \; N(t) = N(0)e^{nt},$$

subject to

$$(2.69) \quad \dot{K} = sY = Y(K, h, t; u) - cN,$$

$$(2.70) \quad Y = AK^{\beta}(huN)^{1-\beta} h_a^{\gamma},$$

$$(2.71) \quad \dot{h} = k(1-u)h,$$

where n, A, γ, k and β are positive constants there being $\beta < 1$ as usual. For $\gamma > 0$ we obviously have increasing returns to the scale.

A new control parameter also appears, viz.u, which is the share of the total working time $N(t)$ used in production in the period of production (i.e. the year) t. The remaining part $(1-u)N$ of the total working time is spent in education and thus devoted to the accumulation of human capital. Learning by doing is not included in the Lucas model. The factor of production huN represents the labour input to production, where the impact of knowledge and skills is taken into account.

The surrogate H^* for the Legendre function G^*, which appears in the Arrow-Kurz generalization, is now given by

$$H^*(K,h,p,q,t;c,u) = N\left(\frac{c^{1-\sigma}-1}{1-\sigma}\right) + p\dot{K}(K,h,t;c,u) + q\dot{h}(h,t;u),$$

where p and q are the Kurz momentums associated with K and h, respectively. The equations of motion accordingly include the following ones:

$$\text{(2.72)} \qquad H_c^* = 0, \quad \text{i.e. } p = c^{-\sigma},$$

$$\text{(2.73)} \qquad H_u^* = 0, \quad \text{i.e. } u = (1-\beta)pY/kqh.$$

The "modified Hamiltonian equations" can then be written for the "current-time Hamiltonian" H^o. They add to the growth equations (69) and (71) the Euler equations

$$\text{(2.74)} \qquad \dot{p}/p = \rho - (1/p)H_K^o = \rho - \beta Y/K$$

$$\text{(2.75)} \qquad \dot{q}/q = \rho - (1/q)H_h^o = \rho - k.$$

The latter is obtained immediately from (73) and from the fact that H^o is equal to H^*, where c and u solve the equations (72) and (73), respectively.

We can see at once that the Legendre condition is satisfied, since in view of (68),(69),(70) and (71) we have:

$$\left\| \begin{matrix} L_{\dot{K}\dot{K}} & L_{\dot{K}\dot{h}} \\ L_{\dot{h}\dot{K}} & L_{\dot{h}\dot{h}} \end{matrix} \right\| = \left\| \begin{matrix} -(\sigma/N)e^{-\rho t}c^{-(\sigma+1)} & 0 \\ 0 & 0 \end{matrix} \right\|.$$

Thus the quadratic form (24) now reduces to

$$-(\sigma/N)e^{-\rho t}c^{-(\sigma+1)}(\dot{K})^2 < 0 \quad \forall t.$$

The natural boundary conditions are

$$L_{\dot{K}} = -e^{-\rho t}c^{-\sigma} \to 0, \quad L_{\dot{h}} \to 0 \quad \text{with } t \to \infty.$$

The first one is satisfied, if $c \to \infty$, which is the case again, at least on the balanced-growth path, of which we are interested. The latter condition is fulfilled, because $L_{\dot{h}} = 0$.

The market clearing then make the exogeneously given and the endogeneously produced levels of human capital to coincide:

$$h(t) = h_a(t) \ \forall t. \tag{2.76}$$

Thus instead of (70) we now have

$$Y = AK^{\beta}(uN)^{1-\beta}h^{1-\beta+\gamma}. \tag{2.77}$$

The transversality conditions in their original form, of course, are the following:

$$\left(\frac{-G}{e^{-\rho t}p}\right)_{t_1} = \left(\frac{\partial_t Y}{Y_K}\right)_{t_1}, \quad \left(\frac{-G}{e^{-\rho t}q}\right)_{t_1} = \left(\frac{\partial_t Y}{Y_h}\right)_{t_1},$$

with $t_1 = \infty$. Here

$$\partial_t Y = n(1-\beta)Y, \ \ Y_K = \beta Y/K, \ \ Y_h = (1-\beta+\kappa)Y/h.$$

But again, as in the Solow model, this original form of transversality conditions would give $-\infty = +\infty$, and it has to be replaced by the form

$$\lim_{t\to\infty} e^{-\rho t}pK = -\frac{\beta}{n(1-\beta)}\lim_{t\to\infty} G, \tag{2.78}$$

$$\lim_{t\to\infty} e^{-\rho t}qh = -\frac{1}{n}\left(\frac{1-\beta+\gamma}{1-\beta}\right)\lim_{t\to\infty} G. \tag{2.79}$$

To see on which conditions they are valid we have to study the solution of the defining equations of motion (69)-(75) of the Lucas growth theory.

Again the balanced-growth path will be the most interesting part of the solution of the equations (69),(71)-(75) and (177), in fact the only part of solution which can be actually worked out. It is defined by

$$Y^* = Y^*(0)e^{\lambda t}, \ \ K^* = K^*(0)e^{\lambda t}, \ \ h^* = h^*(0)e^{\nu t}.$$

This together with (69) gives again

$$c^* = c^*(0)e^{(\lambda - n)t},$$

but for ν and λ we now get, from (71) and (77), respectively, the conditions

$$\begin{aligned} \nu &= k(1-u^*), \quad \text{thus } u^* = \text{Constant}, \\ \lambda &= \beta\lambda + (1-\beta)n + (1-\beta+\gamma)\nu. \end{aligned}$$

Hence three new parameter conditions emerge: first

$$u^* = \frac{k-\nu}{k}, \quad 1-u^* = \frac{\nu}{k}, \tag{2.80}$$

then, because we must have $1 < u^* < 1$,

$$k > \nu > 0, \tag{2.81}$$

and finally:

$$\lambda - n = \left(\frac{1-\beta+\gamma}{1-\beta}\right)\nu > 0. \tag{2.82}$$

With (77) and (80)-(82) the above balanced-growth substitutions solve the growth equations (69) and (71). The balanced-growth equation of physical capital again gives the further relation

$$\lambda = s^* b^*. \tag{2.83}$$

Turning to the Euler equations (74) and (75), the former one together with (72) gives on the balanced-growth path the important parameter condition

$$\rho + \sigma(\lambda - n) = \beta b^*, \tag{2.84}$$

met in the Solow model already. From this together with (83) we have again:

$$\sigma = \frac{\beta b^* - \rho}{s^* b^* - n}, \quad b^* = \frac{n - \rho/\sigma}{s^* - \beta/\sigma}. \tag{2.85}$$

As to the equation (75) it of course is already a general solution and holds good on the balanced-growth path as well.

Of the two conditions (72) and (73) to be imposed on the respective control parameters c and u, the equation (73) gives, as a consequence of a constant balanced-growth share u^* and (84):

$$k = \beta b^* - (\lambda - \nu). \tag{2.86}$$

It remains to consider the transversality conditions (78) and (79). Now the Legendre function G comprises three terms:

$$G = e^{-\rho t} N V(c) + e^{-\rho t} p \dot{K} + e^{-\rho t} q \dot{h}. \tag{2.87}$$

We again have, just like in the Solow model, two possible ways of satisfying the transversality conditions:

Case (i). Here we have the parameter conditions

$$\sigma > 1 \quad \text{and} \quad \rho > n, \tag{2.88}$$

from which, in view of (83) and (85),

$$s^* < \beta \tag{2.89}$$

follows. The conditions (88) make the first term in G to approach asymptotically zero and the condition (89) makes the second term to do so. In order for the third term in G to approach zero asymptotically we need the further parameter condition (81), which of course must be valid for other reasons too. On the conditions (81),(88) and (89) we can thus write the transversality relations in their "textbook form"

$$e^{-\rho t} pK \to 0 \ \text{ and } \ e^{-\rho t} qh \to 0 \ \text{with } t \to \infty. \tag{2.90}$$

In fact, just as in the Solow model, these relations are already satisfied on the same conditions under which they could be written down.

Case (ii). If we reject the condition that $\sigma > 1$, we have to impose on ρ the very strong condition that

$$\rho > \lambda. \tag{2.91}$$

This together with (81) and (89) will in this case allow us to write down the transversality relations in the "textbook form" (90). (And again the same conditions already suffice to satisfy them.)

3 The Canonical Formalism of Macroeconomics

3.1 Generalized Value Function

The Uses of Human Time

Let the total human time in a country within a period of production, say a year t, be defined as the grand total $N(t)$ of the living times of all working-age people in society during the year t:

$$N(t) = \sum_{i=1}^{N^*(t)} N_i(t).$$

Here $N^*(t)$ is the number of grown-up persons, say persons with the age ≥ 16 years, who were alive the whole year t or a part of it, $N_i(t)$ being the living time of the person i in this year. We make the simplifying assumption that

(3.1) $N = N(0)e^{nt}$, with $N(0) = \text{Constant} > 0, n = \text{Constant} > 0.$

The total human time is divided between the leisure $L(t)$ and the working hours $H(t)$, the latter being divided between the hours H_{phys} of physical work in production and the hours H_{int} of intellectual work, this being in turn divided between the hours H_{int}^{paid} of paid intellectual work and the hours H_{ed} spent in formal education as indicated in Fig.1.

The hours of physical work are used up in the production of output. When multiplied by the average skill level or human capital h in the workforce, it gives the labour input or employment E:

(3.2) $$E = hH_{phys}.$$

The hours of intellectual work are used up in the production of human capital, there being

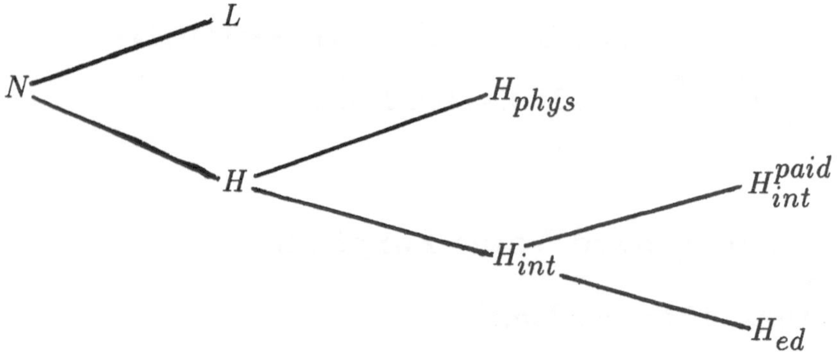

FIG.1. THE USES OF HUMAN TIME

(3.3) $$\frac{\dot{h}}{h} = kH_{int}.$$

Here also the coefficient k is time-dependent and is assumed to grow exponentially so that

(3.4) $$k = k_o e^{mt}, \quad \text{with } k_o = \text{Constant} > 0,\ m = \text{Constant} > 0.$$

Since $k = (\dot{h}/h)/H_{int}$, it is a measure of the (average) speed of learning among those who are doing intellectual working hours.

Who are those people? Everyone who belongs to the active population and works either in the production of output or of human capital, or who is doing hours in formal education, makes a contribution to the hours of intellectual work as measured by H_{int}. The hours of this work thus include, for instance,

- hours of learning by doing while working in the production of output,
- hours of learning spent in formal education,
- hours of teaching in the institutions of formal education,
- hours of planning the work and the tools used up in the production of output,
- hours of planning the teaching in educational institutions, and
- hours of doing managerial work in the production of output or of human capital.

It is evident from this list already that the border line between physical and intellectual work is not observable: you cannot exactly tell, for instance, how much time is used for the learning by doing. It follows that the hours devoted to physical work as well as the hours devoted to intellectual work are not observable variables: both H_{phys} and H_{int} are *hidden variables*, which however are extremely important for the theory.

Let it be emphasized that all the variables discussed in a mathematical theory need not be measurable. It suffices that a theory involving hidden variables predicts correctly a decisive set of observable variables.

Scientific Avantgarde of Technological Progress

We shall write the production function in the form

$$(3.5) \quad Y = AK^{\beta}E^{1-\beta}h_{sc}^{\kappa}, \text{ with } 0 < \beta < 1 \text{ and } 0 < \kappa < 1.$$

where h_{sc} is an exogeneously given factor, which represents the generalized human capital in the sense discussed in Section 1.3. It can be defined as the level of scientific and technological knowledge and skills in society, perhaps even measurable for instance as the level of those knowledge and skills in the scientifically or technologically educated elites in society. But whether the factor h_{sc} is considered as a hidden variable or a measurable one, *it is exogeneously given like the factor representing technological progress in the Solow model.*

It follows that we get the following expression of technological progress in terms of the growth rate of total factor productivity, for which the symbol TFP is here used (note that constant returns to scale are valid in the production function (5)):

$$TFP = \frac{\dot{Y}}{Y} - \beta\frac{\dot{K}}{K} - (1-\beta)\frac{\dot{E}}{E} = \kappa\frac{\dot{h}_{sc}}{h_{sc}}.$$

The technological progress accordingly depends on the growth rate of the general human capital h_{sc} and on the parameter κ, being in fact indicated by their product:

$$(3.6) \qquad TFP = \kappa\frac{\dot{h}_{sc}}{h_{sc}}.$$

Thus the production function (5), together with the interpretation given to the exogeneous factor h_{sc}, is a good expression of the importance of scientific avantgarde of technological progress, as discussed in Section 1.3. The parameter κ obviously indicates how much of the growth rate of h_{sc} is immediately economically profitable and affects the economic growth directly.

The Pursuit of Scientific Progress and Innovations Expands Radically the Macroeconomic Value Function

We have not indicated any particular hours of total human time for the pursuit of scientific and technological innovations. This is partly because innovations cannot be indicated any such particular time. The innovations are not made to order at any specific time but they come when they come, possibly during some hours devoted to some kind of work. But they may be born also during the leisure time. In fact we can expect that most of the scientific and technological innovations start from some ideas, which originally come to a well trained mind during the free wanderings of thought so characteristic of the leisure. Thus we cannot exclude leisure from a scientific formalism discussing the economically relevant use of total human time.

The pursuit of new scientific knowledge certainly implies pursuing nonmaterial values, e.g. the values of objective truth and intellectual beauty, considered as valuable as such. There are other nonmaterial values, for instance beauty as pursued in the fine arts, or a wholistic view of history as pursued in history studies, or a view of the place of man in the world as pursued in philosophy. The original ideas in these fields too may often occur to the mind during leisure. These pursuits also have economic relevance, but on the demand side, as part of economic consumption rather than production: there is a great demand for the works of the fine arts, history and philosophy, but they do not contribute to economic growth in the way science and technology do.

Material and nonmaterial values often mix in human mind, of course, but the 'purely' nonmaterial part of any nonmaterial value – whether the pursuit of truth, beauty or of any view of history or the world for instance – can be represented as economic utility only by a leisure term with an unbounded utility function. Such a leisure term itself can be interpreted to stand for the unbounded value of individual freedom. But it also makes possible those other 'purely' nonmaterial values, which are thus defined as inexhaustible values pursued during the leisure time. (Let it be noted that the utility $V(c)$ of material consumption, given by (2.47), is always finite because of the necessary condition $\sigma > 1$ to be derived in Chapter 5. This expresses the fact that there is an upper limit above which no further material consumption can improve the well-being of an individual.)

It follows that in a correct macroeconomic dynamics, to be constructed in this chapter, the total value function must contain an unbounded leisure term in addition to the usual finite material term. Let two share variables u and v be defined by

$$H = vN, \quad H_{phys} = uH, \quad 0 < u < 1, \quad 0 < v < 1, \tag{3.7}$$

so that the leisure time is expressed as $L = (1 - v)N$. Following the usual praxis in the representation of the material value (see examples in Section 2.2) we can then write the total value function as follows:

$$J = \int_0^\infty e^{-\rho t} \left[N \left(\frac{c^{1-\sigma} - 1}{1 - \sigma} \right) + \xi(1 - v)N \right] dt.$$

Here

$\rho =$ the discount rate,

$\sigma =$ the risk aversion coefficient as used by Lucas (1988),

$c = C/N =$ 'consumption per capita' (here more exactly the consumption per total human time),

$\xi =$ the positive weight of leisure (the wish for individual freedom), to be endogeneously determined as an unbounded function of time.

The maximization of J can be performed by the Arrow-Kurz method, i.e. by maximizing the "current-time Hamiltonian"

$$H^o(K, h, p, q; c, u, v) = N \left(\frac{c^{1-\sigma} - 1}{1 - \sigma} \right) + \xi(1 - v)N + p\dot{K} + q\dot{h},$$

where K, h, p and q are called 'variables', while c, u and v are 'control parameters' and p and q stand for the discounted value prices involved here:

$p(t)$ = the discounted value price of the physical capital K at the future time t,

$q(t)$ = the discounted value price of the human capital h at time t.

The maximization of H^o subject to

$$Y = AK^{\beta}(huvN)^{1-\beta}h_{sc}^{\kappa}, \tag{3.8}$$

$$\dot{K} = sY, \tag{3.9}$$

$$\dot{h}/h = k(1-u)vN, \tag{3.10}$$

where N and k obey (1) and (4), respectively, and s is the net savings rate, gives first:

$$H_c^o = 0, \quad \text{i.e. } p = c^{-\sigma}, \tag{3.11}$$

$$H_u^o = 0, \quad \text{i.e. } u = \frac{(1-\beta)pY}{kqhvN}, \tag{3.12}$$

$$H_v^o = 0, \quad \text{i.e. } v = \frac{(1-\beta)pY}{N[\xi - kqh(1-u)]}, \tag{3.13}$$

after which, i.e. after the parameters c, u and v satisfying (11)-(13) have been inserted in the Hamiltonian H^o, the Euler equations can be written as

$$\text{(3.14)} \quad \dot{p}/p = \rho - (1/p)H_K^o = \rho - \beta Y/K,$$

$$\text{(3.15)} \quad \dot{q}/q = \rho - (1/q)H_h^o = \rho - (1-\beta)pY/qh - k(1-u)vN.$$

The maximization of the generalized value function has now been completed.

The generalized value function can be said to introduce a continuous interaction between material and nonmaterial values as the ultimate source of economic development. For instance the generalized human capital $h_{sc}(t)$, representing the basic nonmaterial value, affects the material output (cf. (8)) and thus the material value function, while the weight $\xi(t)$ of nonmaterial values because of (11) and (20) depends on the material consumption per capita $c(t)$ and thus on the material value function. Also nonmaterial values of the demand side may partake in the interaction.[1]

[1] A detailled discussion of economically relevant nonmaterial values was included in Aulin, The Origins of Economic Growth, Springer-Verlag 1997, Part 1.

3.2 Growth Cycles

Let us introduce the short notations

$$\psi \overset{def}{=} \rho + m - \dot{\xi}/\xi, \tag{3.16}$$

$$\Phi \overset{def}{=} (\dot{\xi}/\xi) + (\dot{\psi}/\psi). \tag{3.17}$$

By dividing the left- and right-hand sides of (12) by the corresponding sides of (13) we get

$$qh = \xi/k. \tag{3.18}$$

From (12) and (15), by using the notation (16) it follows that

$$uvN = \psi/k \tag{3.19}$$

By using (12) together with (18) and (19) we now get

$$pY = \xi\psi/(1-\beta)k. \tag{3.20}$$

It will be shown later that all the variables qh, pY and $uvN = H_{phys}$ decrease exponentially with time. As far as the discounted values qh and pY are concerned this is a triviality: it tells only that the value of human and physical capital, which are realized in a distant enough future, approaches zero for the people living the present time $t = 0$. What is less trivial, but not surprising either, is that according to (19) also the hours of physical work in production decrease exponentially in the future. We shall see later that also the hours of paid intellectual work decrease so that education and leisure increase their significance in total human time (cf. Fig.1).

Cycle Equations: The First Normal Form

From (20), by using (1) for N, (4) for k, (9) for $\dot{K}$, (11) for p and the short notation (17) we get successively:

$$Y = \left[\frac{(1-\beta)k}{\xi\psi}\right]^{\frac{1}{\sigma-1}} \left[\frac{N}{1-s}\right]^{\frac{\sigma}{\sigma-1}}, \tag{3.21}$$

$$\frac{\dot{Y}}{Y} = \left(\frac{\sigma}{\sigma-1}\right)\left(n + \frac{m-\Phi}{\sigma} + \frac{\dot{s}}{1-s}\right). \tag{3.22}$$

By inserting Y from (20) into (14) we have first:

$$K = \frac{Y}{\frac{1}{\beta}\left(\frac{\dot{Y}}{Y} + \rho + m - \Phi\right)}. \tag{3.23}$$

By derivation with respect to time and equating the result with $\dot{K}$ as given by (9) this gives:

$$\frac{\dot{Y}}{Y} - s\frac{Y}{K} = \frac{\frac{d}{dt}\left(\frac{\dot{Y}}{Y} + \rho + m - \Phi\right)}{\frac{\dot{Y}}{Y} + \rho + m - \Phi}.$$

By inserting here $\dot{Y}/Y$ from (22) and Y/K from (23) we get, after some calculation, for the net savings rate s the following second-order differential equation:

$$\frac{d}{dt}\log\left(a\frac{\dot{s}}{1-s} + b\right) = (\beta - s)\left(a\frac{\dot{s}}{1-s} + b\right) - \alpha, \tag{3.24}$$

where

$$a = \frac{\sigma}{\beta(\sigma - 1)}, \tag{3.25}$$

$$b = a\left(\alpha + n - \rho/\sigma\right), \tag{3.26}$$

$$\alpha = \rho + m - \Phi, \tag{3.27}$$

$$\frac{a\dot{s}}{1-s} = x - b, \tag{3.28}$$

the short notation x having been used for the output/capital ratio Y/K.

In terms of the variables s and x we can rewrite (24) in the form of two mutually coupled first-order differential equations, after which the cycle equations (24) on the state-plane (s, x) take the *first normal form*

$$\dot{s} = \frac{1}{a}(1-s)(x-b), \quad \dot{x}/x = (\beta - s)x - \alpha, \tag{3.29}$$

together with (25)-(28). The first equation (29) of course is just a rewritten (28).

The equations (29) are invariant with respect to a linear group of transformations in the space (s, x, t), as will be shown in the following.

Fact 9 and the Continuous Invariance Group

It is easy to see that the equations (29), which because of (25) can be also written as

$$\dot{s} = (\beta - \beta/\sigma)(1 - s)(x - b), \quad \dot{x}/x = (\beta - s)x - \alpha,$$

as well as the equation (28) are invariant with respect to the transformations

$$s \to s' = s, \quad x \to x' = gx, \quad t \to t' = t/g, \quad g > 0 \tag{3.30}$$

of the (s, x, t)-space. These transformations can be interpreted as changes of the time scale, g being a positive scale constant.

For instance, in a transformation with $g > 1$ the real number t, representing in the theory a certain interval of time, obviously becomes smaller because it is divided by g. This is equivalent to the introduction of a larger *time unit* [TU] in this theory. It follows that the output Y', obtained during the new and larger time unit [TU] from a given capital K tied up to production, and thus the new output/capital ratio Y'/K are both of them larger than those obtained from this capital during the old and shorter time unit [TU], in the way expressed by the formulae

$$Y' = gY > Y, \quad x' = Y'/K = gx > x = Y/K, \tag{3.31}$$

respectively.

According to the rules of transformation (30) we have $x \to gx$ and $\dot{s} \to g\dot{s}$, while s is invariant. The parameter b is an output/capital variable and accordingly transforms like x: $b \to b' = gb$. The parameter α also transforms like x. This is because the latter equation (29) tells that

$$\alpha = \dot{x}/x - (\beta - s)x$$

and the variables appearing on the right-hand side of this equation are transformed as follows:

$$\dot{x} \to \dot{x}' = g^2\dot{x}, \quad \dot{x}/x \to g\dot{x}/x, \quad x \to x' = gx,$$

while s and β, being share parameters, are invariant. It follows that

$$\alpha \to \alpha' = g\alpha,$$

and the whole second equation (29) preserves its form:

$$\dot{x}/x = (\beta - s)x - \alpha \longrightarrow \dot{x}'/x' = (\beta - s')x' - \alpha'.$$

The first equation (29) as well preserves its form. This is because 1) $\dot{s}$ transforms like $x - b$, 2) s is invariant, 3) β as a share constant is invariant, and, anticipating a result from Chapter 5, 4) the parameter equation

$$\sigma = \frac{\beta b^* - \rho}{s^* b^* - n}$$

known from the Examples of Chapter 2, is valid also here. The constants ρ and n, defined for the period of production, of course transform as $\rho \to g\rho$ and $n \to gn$, while b^* and s^* transform like b and s, respectively. It follows that the constant σ is invariant in the change (30) of time scale. Thus we have:

$$\dot{s} = \frac{1}{a}(1 - s)(x - b) \longrightarrow \dot{s}' = \frac{1}{a}(1 - s')(x' - b').$$

The equation (28), being identical with the first equation (29), of course also preserves its form.

From the existence of an invariance group involving a transformation of time scale it follows that there is a new degree of freedom to cope with problems of time in the theory in question. We can choose freely the real time interval equivalent to the *theoretical time unit* [TU]. If the real time in years is indicated by τ, a scale constant g_o is determined by the equation

$$T(t)/g_o \text{ [TU]} = T(\tau) \text{ [years]}. \tag{3.32}$$

For instance if, in an application to business cycles, the *average length* of the periods of observed cycles is four years, and the length of period obtained from the theory is $T(t) = 2\pi/\omega$, where ω is the angular velocity of the cycle as seen from the cycle center, we can compute g_o from the equation $g_o = \pi/2\omega$. This scale constant can then be used to cope with the observed average business cycle.

Another important consequence from the possibility to adjust the time scale is that our theory of growth cycles *accounts for the existence of individual cycles of any length.* Thus the fundamental macroeconomic fact 9 as a whole is theoretically reproduced by this theory and thus by the method (b): the stochastic method (a) is not needed for this purpose.

Cycle Equations: The Second Normal Form

Let us consider how the cycle equations of the first normal form behave in the transformation

$$t \longrightarrow -t.$$

Obviously, the equations (28) and (29) are transformed to the equations

$$x - b = -\frac{a\dot{s}}{1-s}, \tag{3.33}$$

$$-\dot{s} = \frac{1}{a}(1-s)(x-b), \quad -\dot{x}/x = (\beta - s)x - \alpha, \tag{3.34}$$

together with (25)-(27), the first equation (34) being a rewritten (33).

One can show that also these equations are competent representations of the second order cycle equation (24). To prove this let (24) be written in the form where the derivation, with respect to time, of the logarithm has been performed:

$$a\ddot{s} + a\frac{(\dot{s})^2}{1-s} = (\beta - s)\,\frac{[a\dot{s} + b(1-s)]^2}{1-s} - \alpha\,[a\dot{s} + b(1-s)] \quad \text{(Original)} \tag{3.35}$$

By applying (28) to the right-hand side the equation assumes the form

$$a\ddot{s} + \frac{a(\dot{s})^2}{1-s} = (1-s)(\beta - s)\,x^2 - (1-s)\,\alpha x. \qquad \text{(Original)} \tag{3.36}$$

We have to compare this second-order equation derived from the original cycle equations (28) and (29) with the corresponding equation derived from the time-inverted equations (33) and (34).

From the first time-inverted equation (34) we first get by derivation with respect to time:

$$-\ddot{s} = \frac{1}{a}[(1-s)\,\dot{x} - (x-b)\,\dot{s}].$$

By applying then the second equation (34) together with (33) this takes the form

$$(3.37)\qquad -a\ddot{s} = (1-s)\,[\alpha x - (\beta - s)\,x^2] + \frac{a(\dot{s})^2}{1-s} \qquad \text{(Time-inverted)}$$

Multiplying both sides of this equation by the factor (-1) and arranging the terms we have:

$$(3.38)\qquad a\ddot{s} + \frac{a\dot{s}^2}{1-s} = (1-s)\,(\beta - s)\,x^2 - (1-s)\,\alpha x. \quad \text{(Time inverted)}$$

But this form of the time-inverted second-order equation is identical with the original second-order equation (36), which proves the invariance of this equation with respect to time inversion.

Thus it has been shown that the equations (28)-(29) can be replaced by the equations (33) and (34) as another possible reduction of the original second-order cycle equation (24) to the normal form. It follows that we have not just one first-order representation of the second-order cycle equation (24) but two of them, given by the equations (28)-(29) and (33)-(34), respectively. Both of them have the same continuous invariance group and thus reproduce the basic fact 9.

The Ordinary Cycles and the Anomalous Cycles

Both acceptable pairs (29) and (34) of the equations of motion of the state $(s(t), x(t))$ of the cycle system on the (s, x)-plane have an obvious special solution, viz. the following one:

$$(3.39)\qquad s = s^* = \text{Constant}, x = b^* = \text{Constant}, \alpha = \alpha^* = (\beta - s^*)b^*.$$

These equations define the cycle center $P = (s^*, b^*)$ as a fixed point on the (s, x)-plane. We can of course accept only such values of the net savings rate s and the output/capital ratio x that obey the conditions

$$(3.40)\qquad s(t) < 1, \quad x(t) > 0 \quad \forall\, t.$$

Under these conditions the second normal form (34) gives:

$$(3.41)\quad \dot{s}\begin{cases} <0 & \text{for } s<1 \text{ and } x>b^*, \\ =0 & \text{for } s=1 \text{ or } x=b^*, \\ >0 & \text{for } s<1 \text{ and } x<b^*. \end{cases} \qquad \dot{x}\begin{cases} <0 & \text{for } x>\dfrac{\alpha^*}{\beta-s}, \\ =0 & \text{for } x=\dfrac{\alpha^*}{\beta-s}, \\ >0 & \text{for } x<\dfrac{\alpha^*}{\beta-s}. \end{cases}$$

This tells that the state (s, x) revolves *counterclockwise* around the cycle center P in the way, which is schematically illustrated in Fig.2. The direction of the flow shown in Fig.2 is horizontal along the curve

$$x = \frac{\alpha^*}{\beta - s}$$

which passes through the cycle center P. The indicated boundaries Γ_1 and Γ_2 of the state space are:

$$\Gamma_1 = \{(s, x);\ s = 1, x > 0\}, \quad \Gamma_2 = \{(s, x);\ s < 1, x = 0\}.$$

If we take the equations of the cycle motion in the first normal form form (29) instead of (34), we get:

$$(3.42)\quad \dot{s}\begin{cases} >0 & \text{for } s<1 \text{ and } x>b^*, \\ =0 & \text{for } s=1 \text{ or } x=b^*, \\ <0 & \text{for } s<1 \text{ and } x<b^*. \end{cases} \qquad \dot{x}\begin{cases} >0 & \text{for } x>\dfrac{\alpha^*}{\beta-s}, \\ =0 & \text{for } x=\dfrac{\alpha^*}{\beta-s}, \\ <0 & \text{for } x<\dfrac{\alpha^*}{\beta-s}. \end{cases}$$

This gives a motion around the cycle center P, which is similar to the motion given by the equations (34), except that now the revolution is *clockwise* – we shall soon study closer these motions.

We shall call the growth cycles with the counterclockwise circulation "ordinary business cycles", and those with the clockwise circulation "anomalous business cycles", because the former case is realized by the cycles around the ordinary balanced-growth path and the latter ones by the cycles around the anomalous growth path mentioned as the fact 16 in Section 1.1.

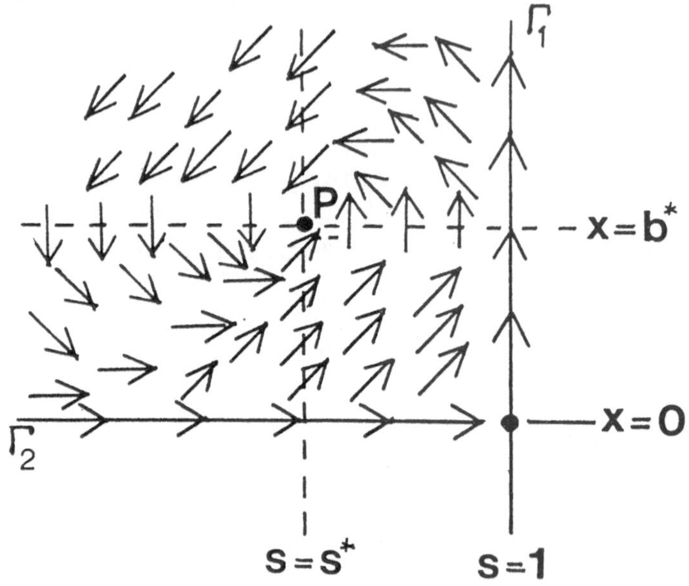

FIGURE 2. – SCHEMATIC ILLUSTRATION OF THE COUNTERCLOCKWISE REVOLUTION OF CYCLES AROUND THE CYCLE CENTER P.

Note. The calculations will be performed by using the linear approximation of cycles in a neighbourhood of the cycle center (as the starting point also in the case of anomalous cycles). The choice of normal form is relevant only in Chapters 5,7 and 9, not elsewhere because of the symmetry of their linear approximations.

Linear Approximations

If the first normal form is used in applications we have to apply the linear approximation of (29) given by

$$\dot{s} = \frac{1-s^*}{a}\,(x-b^*), \tag{3.43}$$

$$\dot{x} = -(b^*)^2\,(s-s^*) + \alpha^*\,(x-b^*). \tag{3.44}$$

In this case a suitable choice of the starting point is:

$$S_+ = [s(0), b^*] \quad \text{where} \quad s(0) > s^*. \tag{3.45}$$

These cycles accordingly start from the right-hand side of the cycle center $P = (s^*, b^*)$ on the line $x = b^*$ and proceeed by revolving the cycle center clockwise.

The solution of (43)-(45) is:

$$s - s^* = e^{\alpha^* t/2}\left(\cos \omega t - \frac{\alpha^*}{2\omega}(\sin \omega t)\right) [s(0) - s^*]_+, \tag{3.46}$$

$$x - b^* = -e^{\alpha^* t/2}\left[\frac{a}{(1 - s^*)}\right] \left[\frac{(\alpha^*)^2 + 4\omega^2}{4\omega}\right] (\sin \omega t)\, [s(0) - s^*]_+. \tag{3.47}$$

Note that

$$[s(0) - s^*]_+ > 0.$$

The angular velocity ω of the flow is given by

$$\omega^2 = (b^*)^2 \left[\left(\frac{1 - s^*}{a}\right) - \left(\frac{\alpha^*}{2b^*}\right)^2\right] \tag{3.48}$$

If the second normal form is chosen, the corresponding linear differential equations approximating (34) are the following ones:

$$\dot{s} = -\frac{1 - s^*}{a}\,(x - b^*), \tag{3.49}$$

$$\dot{x} = (b^*)^2\,(s - s^*) - \alpha^*\,(x - b^*). \tag{3.50}$$

A convenient choice of the starting point of the cycle now is:

$$S_- = [s(0), b^*] \quad \text{where} \quad s(0) < s^*. \tag{3.51}$$

The solution of (49)-(51) is as follows:

$$s - s^* = e^{\alpha^* t/2}\left(\cos \omega t - \frac{\alpha^*}{2\omega}\sin \omega t\right)[s(0) - s^*]_- \tag{3.52}$$

$$x - b^* = e^{\alpha^* t/2}\left[\frac{a}{(1 - s^*)}\right]\left[\frac{(\alpha^*)^2 + 4\omega^2}{4\omega}\right] (\sin \omega t)\, [s(0) - s^*]_-. \tag{3.53}$$

Note that

$$[s(0) - s^*]_- < 0.$$

The angular velocity ω again obeys (48).

The paths defined by (46)-(47) and (52)-(53) on the state-plane (s, x) are symmetric to one another with respect to the axis $s = s^*$, if the starting points S_+ and S_- are situated at the same distance from the cycle center, i.e. if

$$[s(0) - s^*]_- = -[s(0) - s^*]_+.$$

3.3 Solving the Fundamental Equations

General Solution Algorithm

So far in Chapter 3 we have been constructing the fundamental equations of the canonical formalism of macroeconomics. Let us move on to their solution. This corresponds here to the move from phase 1 to phase 2 in the Lucas growth model.

The following solution algorithm makes it possible to calculate the primary variables $Y(t), K(t), h(t), p(t)$ and $q(t)$ as well as the control variables $u(t)$ and $v(t)$ by starting with the known savings rate function $s(t)$. The first five primary functions must be positive for all times, while the share variables u and v must remain between zero and one, and all these functions must be uniquely determined by the constants of the theory and by the functions $\xi(t), \psi(t)$ and $\alpha(t)$ to be determined in Chapter 5. The usual boundary conditions of the canonical formalism must be satisfied.

The positive time functions $N(t)$ and $k(t)$ having been given by (1) and (4), respectively, the following solution algorithm will be given in terms of the time functions $\xi(t), \psi(t)$ and $\alpha(t)$, and of the constants of the theory:

1) We assume the function $s(t)$ to be known.

2) The next step is to find the output function $Y(t)$ from (21).

3) After that the physical capital $K(t)$ is computable from (23) and the value function $p(t)$ from (20).

4) The human capital function $h(t)$ can be obtained from (8), where we now, in phase 2, have to make this function identical with the function $h_{sc}(t)$,

(3.54) $$h(t) = h_{sc}(t) \ \forall t,$$

and uvN has to be written in the form (19). Let it be noted that (54), though formally similar to (2.76) in the Lucas model, has an entirely different content. *The generalized human capital h_{sc} is a nonrival good, different from both the Lucas average rival capital h_a and from the rival goods recommended by Barro and Sala-i-Martin (1995), or by Rebelo (1991), in their linear models. The equation (54) defines a perfect 'knowledge-based society'*, which of course is an idealization of real societies but leads to the theoretical derivation of basic macroeconomic facts.

5) When the human capital function $h(t)$ is known, the value function $q(t)$ of human capital can be computed from (18), and the employment function $E(t)$ from the equation

$$E(t) = h(t)\psi(t)/k(t), \tag{3.55}$$

which is true because of (2), (7) and (19).

6) Finding the function $u(t)$, expressing the share of hours of physical work in total working hours, is now possible from the equation

$$u(t) = \frac{\psi(t)}{\psi(t) + \dot{h}(t)/h(t)}, \tag{3.56}$$

which holds good because of (10) and (19).

7) One can then compute the function $v(t)$ from (19), where the function $u(t)$ is given by (56):

$$v(t) = \frac{\psi(t)}{k(t)u(t)N(t)}. \tag{3.57}$$

Positive Growth Rate of Human Capital over Cycles

The numerical limits $s(t) < 1$ and $x(t) > 0\ \forall t$ must of course be obeyed by the solution $s(t)$. Those limits define the space-plane on which the growth cycles are defined. Assuming that this has been taking care of, the positiveness of the output function $Y(t)$ as determined from (21) is guaranteed, i.e.

$$Y(t) = \left[\frac{(1-\beta)k(t)}{\alpha(t)\xi(t)}\right]^{\frac{1}{1-\sigma}} \left[\frac{N(t)}{1-s(t)}\right]^{\frac{\sigma}{\sigma-1}} > 0 \quad \forall t, \tag{3.58}$$

since also $\xi(t)$ and $\alpha(t)$ will be positive (cf. Chapter 5).

The positivity of the physical capital K,

$$K(t) = Y(t)/x(t) > 0 \quad \forall t,$$

is then clear over all the cycles by the formula (58) and the limit $x(t) > 0$ of the state-plane of growth cycles. The same holds good in the case of the value price $p(t)$, since all the functions in (20) are positive.

The human capital function $h(t)$, as defined in the fourth step of the algorithm, is the function

$$(3.59)\quad h(t) = \left(\frac{k(t)}{\psi(t)}\right)^{\frac{1-\beta}{1-\beta+\kappa}} \left(\frac{Y(t)}{AK(t)^{\beta}}\right)^{\frac{1}{1-\beta+\kappa}} > 0 \quad \forall t,$$

which is positive because all the functions $Y(t), K(t), k(t)$ and $\psi(t)$ are.

The employment function $E(t)$, defined in the fifth step of algorithm by (55), is now of course positive, as well as the value function $q(h)$ of human capital defined by (18) in terms of the positive functions $\xi(t)$ and $k(t)$.

Let us study the condition

$$(3.60)\qquad 0 < u(t) < 1 \ \ \forall t.$$

It follows from (10) and (19) that we can write:

$$(3.61)\qquad u(t) = \frac{\psi(t)}{\psi(t) + \dot{h}(t)/h(t)}.$$

Since $\psi(t)$ is positive, the condition (60) is valid if, and only if

$$(3.62)\qquad \frac{\dot{h}(t)}{h(t)} > 0 \ \ \forall\, t.$$

This implies a condition on the parameter m, as can be shown as follows.

First we get successively, by applying (4),(8),(19) and (54):

$$(3.63)\qquad Y = AK^{\beta}(\psi/k)^{1-\beta} h^{1-\beta+\kappa},$$

$$(3.64)\qquad (1-\beta+\kappa)\frac{\dot{h}}{h} = \frac{\dot{Y}}{Y} - \beta\frac{\dot{K}}{K} + (1-\beta)\left(m - \frac{\dot{\psi}}{\psi}\right).$$

Hence, by the substitutions

$$\frac{\dot{K}}{K} = sx$$

obtained from (9) and

$$\frac{\dot{Y}}{Y} = \frac{\dot{\xi}}{\xi} + \frac{\dot{\psi}}{\psi} - \rho - m + \beta x.$$

obtained from (20) and (14), we get:

$$(3.65) \quad (1 - \beta + \kappa)\,\frac{\dot{h}}{h} = \beta(1 - s)x + (1 - \beta)m - \rho - m + \frac{\dot{\xi}}{\xi} + \beta\frac{\dot{\psi}}{\psi}.$$

Secondly, by means of (16), (17) and (27) we can here write:

$$(3.66) \quad \frac{\dot{\xi}}{\xi} - \rho - m = -\psi,$$

$$(3.67) \quad \frac{\dot{\psi}}{\psi} = \psi - \alpha.$$

This gives the result

$$(3.68) \; (1 - \beta + \kappa)\,\frac{\dot{h}}{h} = \beta(1 - s)\,x + (1 - \beta)\,m - \beta\alpha - (1 - \beta)\psi.$$

In this equation the first term is positive because $s < 1$ and $x > 0$. This leaves us the formula

$$(3.69) \quad (1 - \beta)m > \beta\alpha + (1 - \beta)\psi$$

as the necessary and sufficient condition of (62) and thus of (60).

Let it be noted that since Φ, as defined by (17), may be a function of time and as it appears in the defining formula (27) of the parameter α, the latter too may be a time function. And because α appears in the defining formula (26) of the parameter b, this parameter too may be a function of time. It will be shown in Chapter 5 that ψ and α have a common maximum value:

$$(3.70) \quad \text{Max}\,\psi(t) = \text{Max}\,\alpha(t) = \alpha^* = \text{Constant} > 0.$$

It follows, in view of (69), that the important relation

$$(3.71) \quad m > \frac{\alpha^*}{1 - \beta}$$

is the condition of (62) and thus of (60).

On the other hand (71) is the only condition that must be satisfied in order that the whole algorithm given above is acceptable as a general solution of the present macroeconomic canonical formalism. No problems appear with the corresponding formula $0 < v(t) < 1$ for the allocation variable v, once the initial levels k_0 and $N(0)$ are suitably chosen.

Boundary Conditions

No additional conditions are imposed on the above general solution either by the usual boundary conditions. As far as as the natural boundary conditions and the Legendre condition are concerned, this can be shown in a similar way used in connection of the Lucasian growth mechanics (Section 2.2). There are some differences. Otherwise than there the derivative of the integrand in the value function J, with respect to $\dot{h}$ is not zero in this case, and also the control parameter v, and not only c, is included in the value function. But v is a function of $\dot{h}$, since in view of (10) the equation (13) can be written as

$$v = \frac{(1-\beta)pY}{\xi N - q\dot{h}/v}.$$

Solved for v we get

$$v = \frac{(1-\beta)pY + q\dot{h}}{\xi N}, \quad \text{thus} \ \ v_{\dot{h}} = q/\xi N,$$

which can be shown to produce the desired boundary condition for the derivative of the action function with respect to $\dot{h}$. As to the Legendre condition, it is easily shown to have the same quadratic form as in the Lucas case (for details see Aulin, 1996).

Transversality relations are the only interesting boundary conditions in the present theory, not that they were not satisfied but because of their interesting form. It follows from (4),(19) and (63) that we have now, as a deviation from the Lucasian growth mechanics,

$$\partial_t Y = -(1-\beta)m,$$

while the derivatives Y_K and Y_h have the similar form as in the Lucas case. Thus the transversality conditions of the canonical formalism of macroeconomics will assume the form

$$\lim_{t\to\infty}\left(\frac{-G}{e^{-\rho t}p}\right) = -\left[\frac{m(1-\beta)}{\beta}\right]\lim_{t\to\infty} K, \tag{3.72}$$

$$\lim_{t\to\infty}\left(\frac{-G}{e^{-\rho t}q}\right) = -\left[\frac{m(1-\beta)}{1-\beta+\kappa}\right]\lim_{t\to\infty} h. \tag{3.73}$$

It can be seen from these expressions that the present theory indeed differs very much from the Lucas or Solow models, which in this respect were close to each other. The ultimate reason for this difference is the appearance, in the current-time utility function, of the leisure term with an *unbounded* value function indicating the unbounded value of individual freedom. This affects the character and predictions of economic theory in an essential way, but it also makes simpler the way in which the transversality conditions are satisfied: in the suggested canonical formalism of macroeconomics they are fulfilled in their original Courant-Hilbert form obeyed by the above two equations.

3.4 Cycle Functions Generating the Business Cycles

The growth cycles, as represented by the present minimal macroeconomic dynamics, take place on the plane (s, x). The minimal character of this dynamics is best emphasized just by the fact that only two functions, viz. the net savings rate function $s(t)$ and the output/capital ratio function $x(t)$ suffice to determine the growth cycles of every real economic variable. We need only write for each such variable X its growth rate in the form

$$\frac{\dot{X}}{X} = \left(\frac{\dot{X}}{X}\right)_P + Q_X(s, x), \tag{3.74}$$

where the dependence of the growth rate on the functions s and x is separated from the rest – and we have

1) the growth rate $(\dot{X}/X)_P$ of X at the cycle center P and

2) the *cycle function* $Q_X(s, x)$, which generates the growth cycles known as 'business cycles' of the variable X around the cycle center.

The growth rates $(\dot{X}/X)_P$ are the tools in terms of which the basic macroeconomic facts 1-7 and 16 will be explained in Chapters 5 and 6.

The cycle functions similarly are important tools in the theoretical explanation of the basic facts 10-15 and 17 concerning the business cycles. In fact they are the principal theoretical tool needed for the explanation of those facts, as will be shown by the results of Chapters 7-9, to which the shocks introduced in Chapter 10 add very little. Thus it is mainly the existence of the cycle functions that makes of the present dynamics a minimal dynamics.

Cycle Function of the Output Y

The growth rate of output as a function of x is defined by (3.14) and (3.20), which together give:

$$\frac{\dot{Y}}{Y} = \frac{\dot{\xi}}{\xi} + \frac{\dot{\psi}}{\psi} - \rho - m + \beta x. \tag{3.75}$$

By applying (5.37)-(5.40) this can be written as

$$\frac{\dot{Y}}{Y} = \beta b - \alpha + \beta(x - b). \tag{3.76}$$

Thus we have the following general expression for the growth rate of output at the cycle center P:

$$\left(\frac{\dot{Y}}{Y}\right)_P = \beta b - \alpha. \tag{3.77}$$

The general expression of the cycle function of output is

$$Q_Y = \beta(x - b). \tag{3.78}$$

Cycle Function of the Consumption C

Since $C = (1 - s)Y$ we have first

$$\frac{\dot{C}}{C} = \frac{\dot{Y}}{Y} - \frac{\dot{s}}{1 - s}.$$

Because of the definitions (26) and (27) of the parameters b and α, respectively, that we can write

$$\sigma n + m - \Phi = \sigma(n - \rho/\sigma) + \alpha.$$

With this substitution the equation (22) gives:

$$\left(\frac{\sigma - 1}{\sigma}\right)\frac{\dot{Y}}{Y} = \left[\frac{\sigma(n - \rho/\sigma) + \alpha}{\sigma} + \frac{\dot{s}}{1 - s}\right].$$

It follows that we can write (22) in the following form:

$$\frac{\dot{Y}}{Y} - \frac{\dot{s}}{1 - s} = \frac{1}{\sigma}\frac{\dot{Y}}{Y} + \frac{\sigma(n - \rho/\sigma) + \alpha}{\sigma}.$$

In view of (3.25) and (3.26) the last term is equal to $(1 - 1/\sigma)(\beta b - \alpha)$, and we have the result

$$\begin{aligned} \frac{\dot{C}}{C} &= \frac{\dot{Y}}{Y} - \frac{\dot{s}}{1 - s} = \frac{1}{\sigma}\frac{\dot{Y}}{Y} + (1 - 1/\sigma)(\beta b - \alpha) \\ &= \frac{\beta}{\sigma}(x - b) + \beta b - \alpha. \end{aligned} \tag{3.79}$$

It follows that

$$\left(\frac{\dot{C}}{C}\right)_P = \left(\frac{\dot{Y}}{Y}\right)_P = \beta b - \alpha \tag{3.80}$$

$$Q_C = \frac{Q_Y}{\sigma} = \frac{\beta}{\sigma}(x - b) \tag{3.81}$$

Consumption will accordingly grow at the same rate as output does, and it also has its maximums and minimums during the cycle simultaneously with output. But its standard deviation over a cycle is smaller than that of output (the constant σ proves to be larger than one).

Cycle Function of the Investment I

Since $I = sY$ we have now

$$\frac{\dot{I}}{I} = \frac{\dot{Y}}{Y} + \frac{\dot{s}}{s}.$$

From (79) we get first:

$$\frac{\dot{Y}}{Y} + \frac{\dot{s}}{s} = \left[1 + \left(\frac{1-s}{s}\right)\right]\frac{\dot{Y}}{Y} - \left(\frac{1-s}{s}\right)\left[\frac{\beta}{\sigma}(x-b) + (\beta b - \alpha)\right].$$

This gives, with the substitution (76):

$$\frac{\dot{I}}{I} = \beta b - \alpha + \left[1 + \left(\frac{\sigma - 1}{\sigma}\right)\left(\frac{1-s}{s}\right)\right]\beta(x - b).$$

Thus we get the results

$$\left(\frac{\dot{I}}{I}\right)_P = \left(\frac{\dot{C}}{C}\right)_P = \left(\frac{\dot{Y}}{Y}\right)_P = \beta b - \alpha, \tag{3.82}$$

$$Q_I = \left[\beta + \left(\beta - \frac{\beta}{\sigma}\right)\left(\frac{1-s}{s}\right)\right](x - b). \tag{3.83}$$

In a linear approximation we can put $s = s^*$ to get:

$$Q_I \overset{Lin}{=} \left[\beta + \left(\beta - \frac{\beta}{\sigma}\right)\left(\frac{1-s^*}{s^*}\right)\right](x - b). \tag{3.84}$$

It follows that in the linear approximation output, consumption and investment grow all of them at the same rate and also have their maximums and minimums during a detrended cycle simultaneously. The standard deviation of investment over a detrended cycle will however be remarkably larger than that of output (mainly because of the denominator s^*).

Cycle Function of the Human Capital h

We have already done the groundwork in Section 3.3. By rewriting (68) in the form

$$(3.85)\quad (1-\beta+\kappa)\frac{\dot h}{h} = \beta(1-s^*)\,b + (1-\beta)\,(m-\psi) - \beta\alpha + \\ + \beta[(1-s)\,x - (1-s^*)\,b]$$

we have isolated a function of s and x, which vanishes at the cycle center (s^*, b). Thus we have the result:

$$(3.86)\quad \left(\frac{\dot h}{h}\right)_P = \frac{\beta\,(1-s^*)\,b + (1-\beta)\,(m-\psi) - \beta\alpha}{1-\beta+\kappa},$$

$$(3.87)\quad Q_h = \left(\frac{\beta}{1-\beta+\kappa}\right)\;[(1-s)\,x - (1-s^*)\,b].$$

We can see that both the growth rate and the cycles of human capital differ very much from those of output, consumption and investment. Indeed the real economic variables connected with labour input, viz. human capital, employment and productivity, form another group of growth cycles differing clearly from the output group.

Cycle Function of the Employment E

Since $E = h\psi/k$ we can, in view of (4) and (67), write

$$(3.88)\quad \frac{\dot E}{E} = \frac{\dot h}{h} + \psi - \alpha - m.$$

Substituting (82) and (83) this becomes:

$$\left(\frac{\dot{E}}{E}\right)_P = \frac{\beta\,(1-s^*)\,b + \kappa\,(\psi - m) - (1+\kappa)\,\alpha}{1-\beta+\kappa}, \tag{3.89}$$

$$Q_E = \left(\frac{\beta}{1-\beta+\kappa}\right)[(1-s)\,x - (1-s^*)\,b]. \tag{3.90}$$

Thus employment in a knowledge-based economy analysed in this book has the same cycle as human capital, but it grows at a different rate (much slower than human capital, as will appear later – this theoretical prediction also matches the empirical experience).

Cycle Function of the Productivity Y/E

We can write, because of

$$\frac{(d/dt)(Y/E)}{(Y/E)} = \frac{\dot{Y}}{Y} - \frac{\dot{E}}{E},$$

at once the equations

$$\left[\frac{(d/dt)(Y/E)}{(Y/E)}\right]_P = \left(\frac{\dot{Y}}{Y}\right)_P - \left(\frac{\dot{E}}{E}\right)_P, \tag{3.91}$$

$$Q_{Y/E} = Q_Y - Q_E. \tag{3.92}$$

The differences between the cycle functions of employment and productivity, with respect to the cycle function of output, reproduce theoretically their observed leads and lags in the business cycle, as will be shown in Chapter 7. The cycle functions also explain the other fundamental facts concerning the growth cycles clearly better than do the prevailing RBC theories of business cycles, as we shall see in Chapters 8-10.

The Cycle Functions as Causal Replacements of Slutsky Statistics in Business Cycle Theory

In the 1990s macroeconomic theory has been given a stochastic formulation by a small group of mathematically trained economists. In a recent book some important steps toward the stochastic approach are described as follows[2]:

"Eugen Slutsky (1937), in an important paper[3], put forth an alternative theory [i.e. alternative to a causal approach – A.A.]. He showed how fluctuations resembling business cycles could result from the sum of random shocks to the economy if the economy were characterized by a stable stochastic difference equation with large positive real roots ... Several important developments in growth theory established the foundation that made it possible to think about growth theory and business cycles within the same theoretical framework. One of the most important of these developments from the standpoint of the issues addressed in this book was Brock and Mirman's[4] characterization of optimal growth in an economy with stochastic productivity shocks ... The most thorough and up-to-date treatment of the important theoretical issues in the theory of economic growth is contained in Stokey and Lucas, with Prescott[5]. That theory is the fundamental building block of the modern approach to studying the business cycle."

In a review of Lucas (1987)[6] Richard Day expressed his fundamental criticism[7]:

"Having thus confessed my keen admiration, I turn now to the limitations of this work and the paradigm it offers. I see three that are fundamental. First is the crucial role attributed to shocks. These are impulses, to technology, to tastes, to policy instruments that drive everything of interest. *They* explain why agents have to plan with probabilities in mind, and they ultimately explain the cause for variations in individual behaviour and aggregate measures of economic activity. The implication is that the changes of greatest interest – and that cause

---

[2]T.F.Cooley and E.C.Prescott,Economic growth and business cycles, Cooley (ed.), ibid., p.1-38.

[3]A reference to E. Slutsky, The summation of random causes as the cause of cyclic processes, Econometrica, 1937, p. 105-146.

[4]W.A. Brock and L.J. Mirman, Optimal economic growth and uncertainty: the discounted case, Journal of Economic Theory 4, 1972, p.497-513.

[5]N. Stokey and R.E. Lucas with E.C. Prescott, Recursive Methods in Economic Dynamics, Harvard University Press 1989.

[6]R.E. Lucas Jr., Models of business cycles, Basil Blackwell, London 1987.

[7]Richard H. Day, Review article, Structural Change and Economic Dynamics 3, 1992, p.177-182.

the greatest problems – are explained in terms that are not and cannot be explained. Thus, the Great Depression or the Black Mondays are explained as being caused by shocks to technology, or to preferences or to government error, or to a combination of them. The reader may ask if this is not reintroducing *ad hocery* in the back door after it has been ceremoniously ushered out the front."

What is worse still is that the stochastic approach, based on the superposition of stochastic shocks on microeconomic theory, is entirely out of touch with the nontrivial Plosser fact 7 of growth theory (cf. Chapter 6) and the nontrivial Kydland facts 10-12 about the leads and lags in ordinary business cycles (cf. Chapter 7). It has also been very weak in its attempts to explain the other basic quantitative facts 13-15 concerning the ordinary business cycles, not to speak of the anomalous one.

The stochastic approach will be seen in this book as the last attempt to save the reductionist dogma (cf. Section 1.2) of macroeconomics. Another solution has been suggested in this chapter.

The macroeconomic value function was generalized by adding a leisure term *with an unbounded value function* to the usual consumption term. Such a leisure term can be interpreted to represent the inexhaustible value of individual freedom, which has economic relevance by creating favourable conditions for cognitive and other innovations (cf. Section 3.1).

The cycle functions are the tools through which the fundamental interaction between material and nonmaterial pursuits affects growth cycles in the minimal macroeconomic dynamics constructed in this chapter. The basic quantitative facts 10-15 concerning the ordinary business cycle as well as the existence of the observed anomalous cycle will be theoretically explained in terms of the cycle functions in Chapters 7-10. The stochastic shocks can be superposed on them (Chapter 10), but their effect on the growth cycles proves to be much smaller than what is generally expected by the stochastic school.

Part 2

DERIVATION OF THE 17 BASIC MACROECONOMIC FACTS FROM THE CANONICAL FORMALISM

4 The Principle of Economy in Scientific Explanation

(*A Survey of Empirical Verifications*)

There is a principle of economy of scientific explanation, in economics stressed especially by Milton Friedman, stating that observed facts should be explained by the least possible theoretical assumptions. Following this principle a minimal macroeconomic dynamics, with a generalized value function but without the two dogmas (cf. Section 1.2), was constructed in Chapter 3.

The generalized value function introduced the simultaneous pursuit of both material and nonmaterial values to macroeconomic theory. This was realized in Chapter 3 by introducing a leisure term with an *unbounded* utility function to the macroeconomic value function, in addition to the usual consumption term having a finite utility. Thus nonmaterial values were defined as inexhaustible values pursued outside the time reserved for the pursuit of material values, as represented by working time. Material consumption of course can increase wellbeing only up to a finite limit, while no such limit exists in principle for nonmaterial values such as knowledge or individual freedom.

This term in economic utility leads to the causal method (b) mentioned in Section 1.1 and is, in view of Chapters 5-10, necessary for the theoretical explanation of the basic macroeconomic facts. The usual method (a) based on stochastic shocks superimposed on microeconomic theory fails to explain any nontrivial macroeconomic fact. Its applicability to long-term economic development is questioned.

Nonmaterial values cannot be reduced to microeconomics, because they are products of complex social and intellectual processes, which cannot be reduced to microeconomics. It follows that this theory of macroeconomics cannot be reduced to microeconomics. Thus both two dogmas discussed in Section 1.2 are expressly rejected by the canonical formalism of macroeconomics as constructed in Chapter 3.

We shall discuss in the following separately 1) immediate effects of the generalized value function and 2) the effects derivable from the cycle functions, themselves of course derived from the value function.

Immediate Effects of Generalized Value Function

The appearance of two growth paths:

Chapter 5 shows the two basic growth paths, which exist in the canonical formalism of macroeconomics introduced in Chapter 3. The general solution algorithm of Section 3.3 is applied to two special solutions representing macroeconomic dynamics at the cycle center. The empirical evidence confirming the existence of these two growth paths consists of

– the trivial Kaldor facts 1-6 explained by the balanced-growth path, and of

– the Solow (1957) statistics producing the nontrivial fact 16 explained by the anomalous growth path existing in this theory only.

Section 5.1 gives the first of these special solutions. It is simply a balanced-growth part, whose appearance in every viable growth theory is of course necessary because of the Kaldor facts 1-6. It represents the normal average growth path, whose approximations are usually observed in empirical growth studies. A balanced-growth path therefore is a commonplace in growth theories, and the only interesting points here concern some details of the mathematical representation of economic variables in the macroeconomic dynamics constructed in this book.

Section 5.2 introduces the second special solution, which proves to be an anomalous growth path, whose empirics is described by the nontrivial macroeconomic fact 16 mentioned in Chapter 1. On this growth path the level of output/capital ratio follows a logistically rising curve shown in Fig.3. The appearance of this curve as a consequence of the present theory is an important evidence for the predictive power of this theory. A closer inspection shows that the theory predicts 95 % of the observed higher level .48, to which the level of output/capital ratio rose in the U.S. in the period 1930-48 from its original constant value .30.

The anomalous growth path obviously indicates a disturbance started by the economic crash known as the Great Depression. The empirical change of the constant balanced-growth level .30, valid in the U.S. economy in the period 1909-30, to a logistically rising level valid in the period 1930-48 is exactly described by the Solow (1957) statistics given in Table 1 and illustrated in Fig.3. This evokes two questions discussed at the end of Section 5.2: Why those very clear data have been ignored in economics? How did the output/capital ratio come down from .48 back to its normal level .30, valid in the U.S. economics again in the second half of the century? Only the latter question can be easily answered.

The Plosser fact 7 and other growth effects of saving rate as the first piece of decisive evidence:

Chapter 6 shows that the nontrivial Plosser fact, i.e. a positive effect of savings rate on the growth of output per worker, is theoretically reproduced by the macroeconomic dynamics constructed in Chapter 3. It also quantifies and extends the Plosser fact by showing that changes in the net savings rate produce changes in the growth of output per worker, output and employment, which are verified by economic statistics.

The Plosser fact and its quantification are unique consequences of this particular theory and not likely to appear in any other theory because of certain special conditions it requires. Chapter 6 also shows that the theory works accurately even in a rather complicated and sensitive mathematical environment.

For these reasons the theoretical reproduction and quantification of the Plosser fact 7 must be considered as the first compelling piece of evidence in favour of the macroeconomic dynamics constructed in Chapter 3 and against the usual methods based on microeconomics and stochastic processes.

The Prediction Power of Cycle Functions

The Kydland facts 10-12 on the leads and lags as the second piece of decisive evidence:

Chapter 7 offers another compelling piece of evidence in favour of this dynamics. It gives a theoretical reproduction of the nontrivial Kydland facts 10-12 concerning the observed leads and lags in the business cycle. It is shown that they are correctly predicted by the maximums and minimums of the cycle functions derived in Section 3.4. Two periods in the second half of the century in the U.S. economy both produce similar leads and lags, in empirics as well as theoretically, while the analysed period of the first half of century indicates a system of shorter leads and lags, again backed by the theory as well. These verifications too are quantitative.

Since also the cycle functions are unique characteristics of the macroeconomic dynamics conctructed of Chapter 3, the tests performed in Chapters 6 and 7 can be considered to be the decisive tests of this theory. These results cannot be explained by the stochastic approach.

Note. The Plosser fact and the Kydland facts were only briefly and superficially discussed in my earlier book (Aulin,1997), and the tools used there in these cases were rather ad hoc and even misleading. The more profound discussion given here corrects those errors.

Complementary evidence coming from correlations, standard deviations and autocorrelations over detrended cycles:

Chapter 8 shows that the macroeconomic dynamics constructed in Chapter 3 predicts the correlations with output, the autocorrelations (of output), and the standard deviations of principal real economic variables over detrended ordinary business cycles clearly better than do the principal RBC models of Kydland and Prescott, Hansen and Rogerson, Danthine and Donaldson, and Cooley and Prescott. The ordinary business cycles are cycles around the balanced-growth path.

Chapter 9 shows that also the fact 17 concerning the anomalous correlations over detrended business cycles as observed in the U.S. economy in the period 1914-50 and closely connected with the Great Depression, are theoretically predicted by this theory, which gives their first (roughly) quantitative explanation. They correspond to cycles around the cycle center that moves along the anomalous growth path mentioned above. The anomalous correlations of course are entirely outside the scope of the current stochastic RBC-models.

The role of stochastic optimization in the long-term economic development questioned:

Also the correlations and variances of real economic variables over a detrended cycle are important quantitative data concerning the business cycles. Therefore the results of Chapters 8 and 9 imply a challenge to the method of stochastic optimization, according to which the technological or other shocks are introduced and coupled with economic variables *before* the optimization. The challenge concerns the use of stochastic optimization in the long-term economic development including the business cycles.

Stochastic optimization implies that the economic agents are able to react rationally to the shocks. But there is the other possibility that the agents in general react to trends, observable over some longer interval of time, rather than to the shocks. The above results suggest that *in the long run* they indeed do so: this possibility is realized by the macroeconomic dynamics constructed in Chapter 3, which according to the performed empirical tests predicts the fundamental facts better than do the RBC models based on stochastic optimization.

Chapter 10 tests the mentioned conclusion by superposing technological shocks, of the size usually applied, on the ordinary business cycles as represented by the canonical formalism constructed in Chapter 3. The shocks are thus introduced *after* the maximization of the value function, as perturbations on the strictly causal ordinary cycles given by the maximization of the extended value function. The results suggest that the effect of shocks on correlations and variances over a detrended cycle is small but not negligible. Of the empirical variance of output the causal cycles account for 90 % and the shocks for 10 %.

The results of Chapter 10 tell that the theory of ordinary business cycles given in this book predicts – with and without shocks – the correlations and variances over detrended cycles better than do the conventional RBC models based on stochastic optimization. They follow the patterns of empirical values of correlations and standard deviation proportions rather well, while the stochastic optimization models of Kydland and Prescott, Hansen and Rogerson, Danthine and Donaldson, and Cooley and Prescott follow each other but stray together off from the empirical values.

5 Growth Paths Determined by Canonical Formalism

5.1 The Ordinary Growth Path: Balanced Growth

This is the case where $\dot{s} = 0$ and $\dot{x} = 0$, so that the equations (3.39) are valid. Both pairs of the equations (3.29) and (3.34), and thus the first and second normal forms of cycle equations, are identical in this case.

Solution Algorithm

The general solution algorithm of Section 3.3 gives in this special case successively, after the choices made in the formulae (1) and (2) below:

(5.1) $1 > s = s^* = \text{Constant} > 0, \quad \text{thus } x = b,$

(5.2) $a > 0 \text{ i.e. } \sigma > 1, \; b = b^* = \text{Constant} > 0, \; \psi = \text{Constant},$

(5.3) $\alpha = \alpha^* = (\beta - s^*)b^* = \text{Constant} > 0 \quad i.e. \quad s^* < \beta,$

(5.4) $\psi = \alpha^*, \quad \xi = \xi_o e^{(\rho + m - \alpha^*)t}, \quad \rho + m - \alpha^* > 0, \; \xi_o > 0,$

(5.5) $Y = Y^* = Y_o^* e^{\lambda t}, \; K = K^* = K_o^* e^{\lambda t}, \; h = h^* = h_o^* e^{\nu t},$

(5.6) $\lambda = s^* b^* > 0, \quad \nu = \left(\dfrac{1-\beta}{1-\beta+\kappa}\right)(\lambda + m) > 0,$

(5.7) $v = v^* = v_o^* e^{-(m+n)t}, \quad v_o^* = (\alpha^* + \nu)/k_o N_o < 1,$

(5.8) $1 > u = u^* = \alpha^*/(\alpha^* + \nu) = \text{Constant} > 0,$

(5.9) $p = p^* = p_o^* e^{(\rho - \beta b^*)t}, \quad \text{with } \rho - \beta b^* < 0$

(5.10) $q = q^* = q_o^* e^{(\rho - \alpha^* - \nu)t}, \quad \text{with } \rho - \alpha^* - \nu < 0.$

After the choices (1) and (2) the choices (3)-(10) are uniquely determined. Note that $\rho + m - \alpha^* > 0$ follows from the m-condition (3.71).

Again the Euler equation for the "price" p gives the important parametric relation

$$\rho + \sigma(\lambda - n) = \beta b^*, \tag{5.11}$$

met already in both the Solow and Lucas growth theories. Solved for σ or for b^* this gives, in view of (6),

$$\sigma = \frac{\beta b^* - \rho}{s^* b^* - n}, \quad b^* = \frac{n - \rho/\sigma}{s^* - \beta/\sigma}. \tag{5.12}$$

Important derived variables:

Consumption:

$$C = C^* = C_o^* e^{\lambda t}, \text{ with } C_o^* = (1 - s^*)Y_o^*, \text{thus} \tag{5.13}$$

$$c = c^* = C/N = (1 - s^*)(Y_o^*/N_o)e^{(\lambda - n)t}. \tag{5.14}$$

Investment:

$$I = I^* = I_o^* e^{\lambda t}, \text{ with } I_o^* = s^* Y_o^*. \tag{5.15}$$

Employment:

$$E = E^* = E_o^* e^{(\nu - m)t} \text{ with } E_o^* = \frac{h_o^* \alpha^*}{k_o}. \tag{5.16}$$

Productivity of labour:

$$Y/E = Y^*/E^* = \frac{Y_o}{E_o} e^{(\lambda + m - \nu)t} \tag{5.17}$$

$$= \frac{Y_o}{E_o} \exp\left[\left(\frac{\kappa\nu}{1-\beta}\right) t\right] \tag{5.18}$$

$$= \frac{Y_o}{E_o} \exp\left[\left(\frac{\kappa}{1-\beta+\kappa}\right)(\lambda + m)t\right]. \tag{5.19}$$

Physical capital per worker:

$$K/E = K^*/E^* = \frac{K_o}{E_o} e^{(\lambda + m - \nu)t} \tag{5.20}$$

$$= \frac{K_o}{E_o} \exp\left[\left(\frac{\kappa\nu}{1-\beta}\right) t\right] \tag{5.21}$$

$$= \frac{K_o}{E_o} \exp\left[\left(\frac{\kappa}{1-\beta+\kappa}\right)(\lambda + m)t\right]. \tag{5.22}$$

Productivity of capital, i.e. the output/capital ratio:

(5.23) $$x = Y^*/K^* = b^*.$$

The Kaldor Facts Derived from Theory

Fact 1: It follows from (9) and (11) that

(5.24) $$\lambda > n.$$

Thus per capita output $Y^*(t)/N(t)$ "grows over time", its growth rate $\lambda - n$ being a constant and thus "does not tend to diminish" as required by the Kaldor fact 1.

Fact 2: The physical capital per worker is on the balanced-growth path represented by K^*/E^*, which in view of (21) has the positive rate of growth $\kappa\nu/(1-\beta)$ and thus "grows over time" as required by the Kaldor fact 2.

Fact 3: The rate of return to capital on the balanced growth path is defined by

(5.25) $$\beta(Y^*/K^*) = \beta b^* = \text{Constant} > 0.$$

Thus it is on the balanced growth path not only "nearly constant" as required by the Kaldor fact 3 but constant.

Fact 4: The ratio of output to physical capital is on the balanced growth path defined by $Y^*/K^* = b^*$ and again is not only "nearly constant" as required by the Kaldor fact 4 but constant.

Fact 5: The shares of labour and physical capital in national income are represented by the constants $1-\beta$ and β, respetively, and are constants and not only "nearly constant" as required by the Kaldor fact 5.

Fact 6: It tells that "the growth rate of output per worker differs substantially across countries". The growth rate of output per worker, which on the balanced growth path is given by the growth rate of Y^*/E^*, is according to (18),(3.6) and (3.54) represented by the expressions

(5.26) $$\frac{\kappa\nu}{1-\beta} = \frac{TFP}{1-\beta}.$$

To compare different countries we must here anticipate results from the numerical calculations to be reported in detail in Chapter 6. From the numbers obtained there for the variables $\kappa\nu$ and β we can compute the following estimates for the growth rates of output per worker in the U.S.A, Japan, Canada and France in the Golden Age period, i.e. approximately 1950-73:

USA: 1.88 %, Japan: 5.66 %, Canada: 2.16 %, France: 3.87 %.

Thus the growth rates of output per worker, estimated on the basis of the macroeconomic dynamics constructed in Chapter 3, indeed differ greatly from a country to another. This growth rate was in Japan three times and in France two times that of the U.S.A. Only in Canada it was close to the value of the U.S.A., being only slightly larger than that. The Kaldor fact 6 too is accordingly well reproduced by this theory.

The Boundary Conditions

The natural boundary conditions are satisfied because of (9) and (10). As to the transversality conditions (3.72)-(3.73), we can see at once that the right-hand sides go on the balanced-growth path to negative infinity with $t \to \infty$.

Consider then the left-hand sides

$$-G/e^{-\rho t}p^* = -G^*/p^* \quad \text{and} \tag{5.27}$$

$$-G/e^{-\rho t}q^* = -G^*/q^*, \quad \text{where} \tag{5.28}$$

$$G^* = NV + \xi(1-v)N + p^*\dot{K}^* + q^*\dot{h}^*, \quad V = \frac{c^{1-\sigma}-1}{1-\sigma} \tag{5.29}$$

We have to study study the behaviour of each term when $t \to \infty$. There is now, because of (2), $\sigma > 1$, in which case the function $V(c)$ approaches asymptotically a positive constant. The absolute values of the two first terms in G^*/p^* go to infinity with $t \to \infty$ as the following two exponential factors:

$$NV/p^* \sim e^{[n+\sigma(\lambda-n)]t}, \quad \xi(1-v)N/p^* \sim e^{(m+n+\lambda)t},$$

the first term because of (3.1),(9),(11) and (24), the second one because of (3.1), (4), (9), (11) and (24).

The absolute values of the third and fourth terms go to infinity in a similar way as shown by the formulae

$$\dot{K}^* \sim e^{\lambda t}, \quad q^* \dot{h}^*/p^* \sim e^{\lambda t},$$

the former one because of (5) and the latter one because of (4)-(6),(9) and (10). Thus also the left-hand side of the transversality condition (3.72) goes to negative infinity with $t \to \infty$, which means that this condition is satisfied in its original Courant-Hilbert form discussed in Chapter 2.

Consider then the left-hand side (28), $-G^*/q^*$, i.e. the transversality condition on human capital. The absolute value NV/q^* of the first term behaves for $t \to \infty$, because of (10), like the following exponential function:

$$NV/q^* \sim e^{(\alpha^* + \nu + n - \rho)t},$$

For $\rho < \alpha^* + \nu + n$ this goes to infinity, otherwise it approaches asymptotically a constant, possibly zero. Its behaviour is not important for the validity of transversality conditions. This is because the absolute values of the other terms in (28) always rise to infinity, as shown by

$$\xi(1-v)N/q^* \sim e^{(m+n+\nu)t}, \quad p^*\dot{K}/q^* \sim e^{\nu t}, \quad \dot{h}^* \sim e^{\nu t},$$

the first of them because of (3.1),(4) and (10), the second one because of (4),(5),(9) and (10), and the third one because of (5).

It follows that also the second transversality condition (3.73) is fulfilled on this path of the generalized dynamics. Both transversality conditions are now satisfied in their original Courant-Hilbert forms studied in Chapter 2.

Let it be remarked that we also have the formulae

$$e^{-\rho t} p^* K^* \sim e^{-\alpha^* t} \to 0, \text{ when } t \to \infty, \tag{5.30}$$

$$e^{-\rho t} q^* h^* \sim e^{-\alpha^* t} \to 0 \text{ when } t \to \infty, \tag{5.31}$$

the first one because of (3)-(5) and (9), the second one because of (5) and (10). This time, however, these formulae do not suffice to guarantee the validity of transversality conditions.

5.2. Anomalous Growth Path After a Crash

The balanced-growth path of course is the ordinary case, in terms of which a large mass of empirical data has been analysed in growth theory, and the present theory is no exception in this respect. Chapter 6 will give a nontrivial example of the analysis of empirical data in terms of the balanced growth path as derived from the canonical formalism of macroeconomics constructed in Chapter 3. But there is also an anomalous growth path in this theory, which accounts for the anomalous average growth observed after the Great Depression.

Solution Algorithm: Moving Cycle Center

The anomalous growth path is obtained by starting the general solution algorithm with the following choices:

(5.32) $1 > s = s^* = \text{Constant} > 0, \quad \text{thus } w = b,$

(5.33) $a > 0 \text{ i.e. } \sigma > 1, \quad b \neq \text{Constant}, \quad \psi \neq \text{Constant}.$

Now the cycle center P is not fixed but moves, because of $b \neq$ Constant, along the axis $s = s^*$ on the (s, x)-plane (see Fig.1). We shall study the anomalous case in which the equations (3.28) and (3.29) apply and must be used to determine the growth path. From (3.29) we can see that the parameter b has now to obey the equation

$$\dot{b}/b = (\beta - s^*)b - \alpha.$$

In view of (3.26) this is equivalent to

$$\dot{b}/b = (n - \sigma/\rho) - (s^* - \beta/\sigma)b.$$

Here we can express $n - \rho/\sigma$ in terms of the value $b = b^* > 0$ pertaining to a balanced-growth path, obeying the equation (12) above, to get:

(5.34) $$\dot{b}/b = (s^* - \beta/\sigma)(b^* - b).$$

This is solved by

(5.35) $$b = \frac{b^*}{1 + Be^{-\gamma t}}, \quad \text{where } \gamma = n - \rho/\sigma, \quad B = \frac{b^* - b(0)}{b(0)},$$

which defines a logistic function with two branches.

For $\gamma < 0$ it gives a downward curve to which we shall shortly return at the end of this chapter. The other option $\gamma > 0$ gives the branch of the logistic function, which rises asymptotically toward the balanced growth path with a constant $x = b^*$. This choice corresponds to a cycle center rising toward the value $x = b^*$ on the (s, x)-plane. It adds to the conditions (32) and (33) the further parameter conditions

$$(5.36) \qquad \dot{b} > 0, \quad \gamma > 0 \quad \text{i.e. } s^* - \beta/\sigma > 0.$$

The last condition is in view of (12) equivalent to the second one, since b^* is positive.

In the real world there is no asymptotic approach that would go on and on without limit. Therefore the solution $b(t)$ determined by (35) and (36) suggests that the anomalous growth path is temporary and sooner or later goes over to a balanced-growth path. This indeed proves to be true in the remarkable case, in which reliable empirical statistics is available and the existence of the anomalous growth path can be verified, as we shall see.

Next consider the parameter α. Because of (3.26) also α is a function of time defined by the time function $b(t)$. By applying (3.25) and (3.26), and for γ again (35), we get

$$(5.37) \qquad \alpha = (\beta - \beta/\sigma)b - \gamma = \alpha^* - (\beta - \beta/\sigma)(b^* - b),$$

which expresses the relation between the time-dependent parameter α and the corresponding balanced-growth parameter α^*.

For the auxiliary ψ we now get, with the help of (3.16),(3.17) and (3.27), the differential equation

$$(5.38) \qquad \dot{\psi}/\psi = \psi - \alpha.$$

This is solved, in view of (34) and (37), by

$$(5.39) \qquad \psi = (\beta - s^*)b > 0 \;\forall t.$$

This gives, in view of (3.16):

$$(5.40) \qquad \xi = \xi_o e^{(\rho + m)t - \int_0^t \psi\, dt}, \quad \xi_o > 0.$$

Thus we have settled the two first steps in the general solution algorithm. After the choices (32), (33) and (36) have been made, the anomalous growth path is uniquely determined.

The basic theoretical variables:

For the next steps we first get from (3.22), by applying $\dot{s} = 0$, (3.27) and the above formulae (37) and (6) in this order,

$$\dot{Y}^\dagger / Y^\dagger = \lambda - (\beta/\sigma)(b^* - b), \tag{5.41}$$

where λ is the balanced-growth rate of output. This gives

$$Y^\dagger = Y_o^\dagger e^{\lambda t - (\beta/\sigma)\left(b^* t - \int_o^t b dt\right)}, \quad \text{and } K^\dagger = Y^\dagger / b, \tag{5.42}$$

which settles the third and fourth steps of the algorithm.

To do the fifth step we first calculate, with the help of the production function (3.63) and the above formulae (34),(37),(39) and (41),

$$\frac{\dot{h}^\dagger}{h^\dagger} = \nu + \left(\frac{\beta}{1 - \beta + \kappa}\right)(s^* - \beta/\sigma)(b^* - b), \tag{5.43}$$

where ν is the growth rate of human capital on the balanced growth path. This gives:

$$h^\dagger = h_o^\dagger \exp\left\{\nu t + \left(\frac{\beta}{1 - \beta + \kappa}\right)\left[\gamma t - (\gamma/b^*)\int_0^t b\, dt\right]\right\}. \tag{5.44}$$

To get the sixth step done we can now write directly, in view of (3.56),

$$u^\dagger = \frac{\psi}{\psi + \dot{h}^\dagger / h^\dagger}, \tag{5.45}$$

where $\dot{h}^\dagger / h^\dagger$ is given by (43) and ψ by (39).

The seventh step is equally simple, because (3.57) now gives

$$v^\dagger = \frac{1}{kN}\left(\frac{\dot{h}^\dagger}{h^\dagger} + \psi\right), \tag{5.46}$$

which completes the construction of the basic theoretical variables.

It remains to study the Euler equations (3.14) and (3.15). They give:

$$\dot{p}^\dagger/p^\dagger = \rho - \beta b, \quad \dot{q}^\dagger/q^\dagger = \rho - \psi - \dot{h}^\dagger/h^\dagger. \tag{5.47}$$

Here b is to be taken from (35), ψ from (39) and $\dot{h}^\dagger/h^\dagger$ from (43), after which the function $q^\dagger(t)$ can be calculated in a way similar to the calculations of $Y^\dagger$ in (42) and $h^\dagger$ in (44).

The natural boundary conditions are satisfied, because it follows from (47) that $e^{-\rho t}p^\dagger$ and $e^{-\rho t}q^\dagger$ both approach asymptotically zero. The Legendre condition of course holds good generally in the dynamics of Chapter 3, as was mentioned there.

Since it has been already proved that the transversality conditions (3.72) and (3.73) are satisfied on the balanced-growth path, the case of the anomalous path is trivial. This is because the logistic growth path, studied in this section, asymptotically approaches a balanced-growth path, as is evident from (35),(37), (39),(40),(41), (43),(45),(46) and (47). Thus we have already proved the validity of transversality conditions also in the case of the logistic growth path.

The functions $e^{-\rho t}pK$ and $e^{-\rho t}qh$ vanish asymptotically also on the logistic growth path, but this does not guarantee the validity of the transversality conditions, which are also in this case valid in their original Courant-Hilbert form derived in Chapter 2.

Verification by the Solow (1957) Statistics

At the very beginning of the modern era of growth theory, in Solow's seminal works in the 1950s already, an empirical material was published which leaves no doubt about the existence of the anomalous growth path, on equal justification with the normal balanced-growth path. This material in Solow's 1957 paper was then, of course, published to serve another purpose, viz. that of supporting Solow's representation of technological progress by the exponential function $A(t)$ in his production function.

Finding statistical materials useful for the verification of a general theory was then, and still is, a demanding task if only because economic statistics normally has been collected mainly for variable practical uses and keeps rarely the same standards and units over any longer period.

Solow wrote[1] "The capital time series is the one that will really drive a purist mad. For present purposes, 'capital' includes land, mineral deposits, etc. Naturally I have used Goldsmith's estimates (with government, agricultural, and consumer durables eliminated). Ideally what one would like to measure is the annual flow of capital services. Instead one must be content with a less utopian estimate of the stock of capital goods in existence... Lacking any reliable year-to-year measure of the utilization of capital I have simply reduced the Goldsmith figures by the fraction of the labor force unemployed in each year, thus assuming that labor and capital always suffer unemployment to the same percentage."

TABLE 1. – THE SOLOW DATA FOR CALCULATION OF THE OUTPUT/CAPITAL RATIO*

Year	Priv.nonfarm GNP per manhour (1)	Employed capital per manhour (2)	Output per capital (3)	Year	Priv.nonfarm GNP per manhour (1)	Employed capital per manhour (2)	Output per capital (3)
1909	.623	2.06	.302	1930	.880	3.06	.288
1910	.616	2.10	.293	1931	.904	3.33	.271
1911	.647	2.17	.298	1932	.897	3.28	.273
1912	.652	2.21	.295	1933	.869	3.10	.280
1913	.680	2.23	.305	1934	.921	3.00	.307
1914	.682	2.20	.310	1935	.943	2.87	.329
1915	.669	2.26	.296	1936	.982	2.72	.361
1916	.700	2.34	.299	1937	.971	2.71	.358
1917	.679	2.21	.307	1938	1.000	2.78	.360
1918	.729	2.22	.328	1939	1.034	2.66	.389
1919	.767	2.47	.311	1940	1.082	2.63	.411
1920	.721	2.58	.279	1941	1.122	2.58	.435
1921	.770	2.55	.302	1942	1.136	2.64	.430
1922	.788	2.49	.316	1943	1.180	2.62	.450
1923	.809	2.61	.310	1944	1.265	2.63	.481
1924	.836	2.74	.305	1945	1.296	2.66	.487
1925	.872	2.81	.310	1946	1.215	2.50	.486
1926	.869	2.87	.303	1947	1.194	2.50	.478
1927	.871	2.93	.297	1948	1.221	2.55	.479
1928	.874	3.02	.289	1949	1.275	2.70	.472
1929	.895	3.06	.292				

* Numbers in columns (1) and (2) are given in 1939 dollars.

The first and second columns of Table 1 were composed and published as the fifth and sixth column, respectively, in the corresponding Table 1 of Solow (ibid., p.315). By dividing the numbers in the column

[1] R.M. Solow, Technical change and the aggregate production function, The Review of Economics and Statistics, 1957, p.312-320.

of "Private nonfarm GNP per manhour" by the corresponding numbers in the column "Employed capital per manhour" (the titles used by Solow), one gets the third column of our Table 1. It gives estimates of the annual output/capital ratios Y/K in the U.S. economy from 1909 to 1949, and their graphical illustration is given in Fig.3.

The business cycles in employment are seen as the oscillations of the variable Y/K in the picture. What is interesting for the present purpose, however, is the behaviour of the level of Y/K, which is to be compared with the theoretical level as represented by the variable b in the generalized dynamics. The numbers and the picture suggest that there was indeed a clear-cut balanced-growth path, i.e. a constant $b = b^*$ from 1909 until the Great Depression of 1929-33. But after that there was an equally clear-cut rise of the logistic type in the level of the output/capital ratio that lasted some twenty years, approaching again a constant but higher value toward the end of 1940s.

Thus the qualitative behaviour of the U.S. economy in the period 1909-30 in terms of the present theory corresponds to the balanced-growth path, while the period 1930-48 suggests the anomalous growth path, with its logistic rise in the level of the output/capital ratio toward a new balanced-growth path. The lines drawn in Fig.3 approximately represent the constancy of b prevailing before the Great Depression and the approximately logistic rise during 1930-48. This suggests that anomalous growth path is temporary and appears after an economic crash, going sooner or later over to the balanced-growth path.

The appearance in the Solow statistics of the two different kinds of average growth paths, predicted by the macroeconomic dynamics constructed in Chapter 3, no doubt is a significant fact in favour of this theory: For instance the Solow growth model and the Lucas 1988 economic mechanics have both of them only the balanced-growth path.

So far the verification of the growth types has been only qualitative. Can we bolster it by a correct quantitative prediction of the observed rise of the output/capital ratio from its level $b^* \approx .30$ in the period 1909-1930 to the higher level in 1948 approximating the asymptotic value $b^* \approx .50$?

By applying to the pre-1930 period the same parameter values, which will be used later in Chapter 6 (in the 'historical calibration'

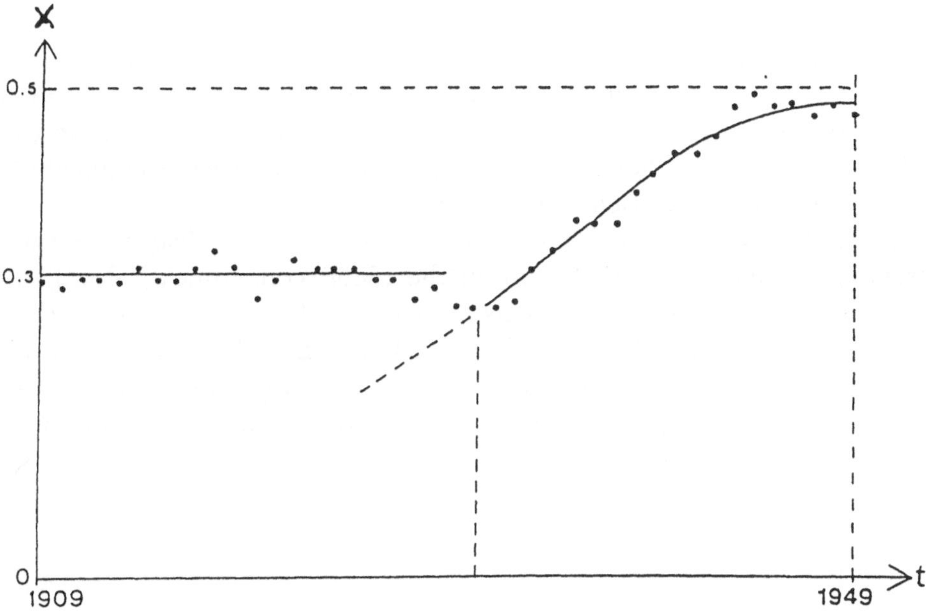

FIGURE 3. – THE TWO AVERAGE GROWTH PATHS VERIFIED BY THE SOLOW (1957) DATA

referring to the first half of the century) and in Chapter 7 ($s^* = .13$), i.e.

$$n = .015, \quad \rho = .02, \quad \sigma = 2.2917, \quad T = 60.707103,$$

where T is the period of the business cycle in theoretical units and corresponds to 4 years, we get:

$$(5.48) \quad b(5T) = \frac{.5}{1 + \left(\frac{.5 - .3}{.3}\right) e^{-(.015 - .02/2.2917)(303.53551)}}$$

$$= (.909966)(.5) \approx .455.$$

Thus the theory explains 90% of the assumed asymptotic value .50 and 95% of the observed value .48 of Y/K toward the end of 1940s, which must be considered as a good verification of the anomalous growth path predicted by the canonical formalism of macroeconomics constructed in Chapter 3.

If the Solow data support the existence of the two average growth paths so clearly, two questions arise. First of all: why this evidence has been ignored ever since 1957 when the Solow data were first published? This is really hard to answer. Maybe even the scientists are prone to look elsewhere, when they encounter phenomena that are not explicable by the prevalent theoretical orthodoxy?

Secondly: How did the arisen level .48 of the output/capital ratio, observed by Solow toward the end of the 1940s, come quickly down to its old constant value of .30, registered in the U.S.A. already as the average of the Golden Age period 1947-73 (see the calibration given in Chapter 6)? The answer offered by the present theory of course refers to the case $\gamma < 0$ mentioned above, but no exact quantitative data are in this case available.

6 The Plosser Fact and Other Growth Effects of Savings Rate

(*The first decisive piece of evidence*)

We shall study in this chapter how the generalized value function affects the growth of variables connected to the labour input E on the balanced-growth path. The Plosser fact 7 states the positive effect of savings rate on the growth of output per worker Y/E, while no similar effect exists in the case of output or employment. This will be theoretically reproduced, and the effects of savings rate on the growth of output and employment will be analysed. The Plosser effect, with a positive effect on Y/E and an absence of it for Y or E, is a puzzle in growth theories. The solution given here is based on the extraordinary connections, which in this theory exists between the net savings rate and the level of output per worker, output and employment.

6.1 The Extended Growth Effects of Savings Rate Following from the Extended Value Function

The mentioned connections derive from the formula (3.19) telling that the hours of physical work in production uvN are, because of the extended value function, represented by ψ/k. On the balanced-growth path ψ is in view of (5.4) and (5.2) equal to the constant

(6.1) $$\psi = \alpha^* = (\beta - s^*)b^*.$$

Since the labour input or employment is in this theory, just as in the Lucas growth model, equal to the average knowledge and skills in the workforce h times the hours of physical work in production, we have

(6.2) $$E^*(t) = h^*(t)\alpha^*/k(t).$$

But the parameter α^* and therefore the employment E^* depend on the net savings rate s^*. This dependence is based on the equation (1) and on the balanced growth Euler equation (5.11), i.e.

(6.3) $$b^* = \frac{n - \rho/\sigma}{s^* - \beta/\sigma} > 0.$$

It follows from (1)-(3) that

$$\frac{\partial \alpha^*}{\partial s^*} = -b^* \left(\frac{\beta - \beta/\sigma}{s^* - \beta/\sigma} \right) < 0, \tag{6.4}$$

provided that $s^* - \beta/\sigma > 0$ as it usually is.

Twofold Dependence on Savings Rate

It follows from (1)-(4) and from the balanced-growth expressions of output and productivity, obtained from (3.63),(5.5) and (5.17), that

$$E^* = \left[\frac{h_o^* \alpha^*}{k_o} \right] \cdot e^{(\nu - m)t}, \tag{6.5}$$

$$Y^* = A[K_o^*]^\beta [\alpha^*/k_o]^{1-\beta} [h_o^*]^{1-\beta+\kappa} \cdot e^{\lambda t}, \tag{6.6}$$

$$\frac{Y^*}{E^*} = \frac{A[K_o^*]^\beta k_o^\beta}{[h_o^*]^{\kappa - \beta} (\alpha^*)^\beta} \cdot e^{(\lambda + m - \nu)t} \tag{6.7}$$

Obviously, when the net savings rate s^* rises, the level $Y^*(0)/E^*(0)$ of output per worker (productivity) also rises, while the levels $Y^*(0)$ and $E^*(0)$ of output and employment, respectively, fall.

But there is also the ordinary dependence on the savings rate of the growth rate λ of output and, in this theory, also that of the growth rate ν of human capital. From (5.6) and (3) we get successively:

$$\lambda(s^*) = s^* \left(\frac{n - \rho/\sigma}{s^* - \beta/\sigma} \right), \tag{6.8}$$

$$\nu(s^*) = \left(\frac{1-\beta}{1-\beta+\kappa} \right) [\lambda(s^*) + m] \tag{6.9}$$

$$\frac{\partial \lambda}{\partial s^*} = -\left(\frac{\beta}{\sigma} \right) \left(\frac{b^*}{s^* - \beta/\sigma} \right), \tag{6.10}$$

$$\frac{\partial \nu}{\partial s^*} = \left(\frac{1-\beta}{1-\beta+\kappa} \right) \frac{\partial \lambda}{\partial s^*}. \tag{6.11}$$

What we need of course is the total dependence of the output per worker Y^*/E^*, the employment E^* and the output Y^* on the savings rate s^*. In empirical verifications of the growth effects of savings rate the effects on the level factors (5)-(7) prove to be decisive as explanations of the observed facts. This explains why the ordinary growth theories, where no level effects appear, fail to predict the Plosser fact and other observed effects of savings rate. But first we need a method that suits for an analysis of empirical observations.

The Differential Method

From (5)-(11) we obtain after some calculation:

$$(6.12)\quad \frac{\partial(Y^*/E^*)/\partial s^*}{Y^*/E^*} = \frac{\beta b^*}{\sigma(\alpha^*)(s^* - \beta/\sigma)}\left[\beta(\sigma-1) - \alpha^*\left(\frac{\kappa}{1-\beta+\kappa}\right)^2(\lambda+m)\right],$$

$$(6.13)\quad \frac{\partial E^*/\partial s^*}{E^*} = (\nu - m)\frac{\partial \nu}{\partial s^*} + \frac{\partial \alpha^*/\partial s^*}{\alpha^*},$$

$$(6.14)\quad \frac{\partial Y^*/\partial s^*}{Y^*} = \lambda\frac{\partial \lambda}{\partial s^*} + (1-\beta)\frac{\partial \alpha^*/\partial s^*}{\alpha^*}.$$

The Plosser fact reproduced by theory:

This is the easy part of the task. The positive Plosser fact is theoretically reproduced as soon as

$$(6.15)\quad \frac{\partial(Y^*/E^*)/\partial s^*}{Y^*/E^*} > 0$$

holds good when the numerical value of the function (12) is computed from the empirical estimates of the parameters appearing in this function.

This will be shown to be the case in all the five cases of verification, involving four countries, to be analysed in this chapter. Let it be noted that it is only the twofold dependence of the relevant economic variables on the savings rate that makes possible the positive Posser effect as well as its quantifications and extensions.

Quantitative prediction of the growth effects of savings rate:

Given two periods of time called BASIC PERIOD 1 and PREDICTED PERIOD 2, and the empirical values of the growth rates of output Y, productivity of labour Y/E and employment E in these two periods in a chosen country, i.e.

$$\lambda_1, \quad (\lambda - \dot{E}/E)_1, \quad (\dot{E}/E)_1, \quad \text{and}$$
$$\lambda_2, \quad (\lambda - \dot{E}/E)_2, \quad (\dot{E}/E)_2,$$

respectively, we can apply the differential method to the prediction of other two observed changes from the observed change of one of these variables in the following way.

If, say, the observed change of productivity, i.e. output per worker

$$\Delta(\lambda - \dot{E}/E)_{obs} = (\lambda - \dot{E}/E)_2 - (\lambda - \dot{E}/E)_1,$$

is assumed to be perfectly predicted by the theory, the theoretical predictions for the other observed changes

$$\Delta\lambda_{obs} = \lambda_2 - \lambda_1 \quad \text{and}$$
$$\Delta(\dot{E}/E)_{obs} = (\dot{E}/E)_2 - (\dot{E}/E)_1$$

are given by

$$\text{(6.16)} \quad \Delta\lambda_{pred} = \frac{(\partial Y^*/\partial s^*)/Y^*}{[\partial(Y^*/E^*)/\partial s^*]/(Y^*/E^*)}\Delta(\lambda - \dot{E}/E)_{obs},$$

$$\text{(6.17)} \quad \Delta(\dot{E}/E)_{pred} = \frac{(\partial E^*/\partial s^*)/E^*}{[\partial(Y^*/E^*)/\partial s^*]/(Y^*/E^*)}\Delta(\lambda - \dot{E}/E)_{obs}.$$

Then the ratios

$$\text{(6.18)} \qquad \frac{\Delta\lambda_{pred}}{\Delta\lambda_{obs}} \quad \text{and} \quad \frac{\Delta(\dot{E}/E)_{pred}}{\Delta(\dot{E}/E)_{obs}}$$

tell how much of the observed changes in the growth rates of output and employment, respectively, are predicted by the theory.

6.2 Empirical Verifications of the Extended Growth Effects of Savings Rate in Various Countries

Problems of Economic Statistics

We have to study the growth rates of output per worker, output and employment in various countries in order to verify the existence of the Plosser fact and other extended growth effects of savings rate. Here we encounter several problems of measurement, because the production of accurate empirical estimates of growth rates is not an easy problem of economic statistics. To guarantee an acceptable level of accuracy some rules have to be observed.

First of all, periods must be rather long in order to avoid errors in empirical estimates due to irregular movements of economy. Secondly, the periods chosen for the verification should be composed of a number of full business cycles: the final year of each period should indicate the same phase of the cycle, in which the period started. To a certain extent this can be tested by varying the first and last years of each period by 1-2 years and repeating the computation in each case to check, whether the result is essentially affected or not. Indeed even a change of one single year in the considered period may remarkably change the empirical estimates obtained from growth statistics, just because of the business cycles.

Thirdly, national economies are not closed systems but interact in many ways with the external world. The fundamental theory on the other hand must start with the study of closed systems in order to reveal the internal mechanism of national economy. The present theory is of course a case in point. Therefore only big nations should be considered as reliable sources of empirical information, since economies in such countries probably are less influenced by external factors. But there have been some great external shocks, which have affected big countries as well. The greatest of the shocks of this kind during the time we are going to study were the sudden remarkable raises in the oil prices of the OPEC countries in 1970s. As a consequence, 1974-82 was an irregular period in most Western economies.

The irregular period 1974-82 splits the modern economic era, say from 1947 to 1990, in two pieces leaving only the years 1947-73 and

1983-90 free from larger irregularities. But the latter period is in itself too short. The impact of these facts on the present study is that we have to make our BASIC PERIODs and PREDICTED PERIODs to overlap in order to get long enough periods for the analysis: both of these periods must be taken from the Golden Age period 1947-73. As a consequence the observed changes in the aggregate economic variables are rather small. On the other hand this makes of the following empirical verifications rather strict and conclusive: they indicate that this theory of the growth effects of savings rate works exactly even in a complicated and sensitive mathematical environment.

The basic source of empirical estimates used in these verifications has been

– R. Summers and A. Heston, The Penn World Table (Mark 5). These tables are published since 1991 and contain yearly information about the economies of 150 countries since 1950. The numbers computed from the Penn Table are marked PENN in the following.

The following other sources of empirical estimates are used:

– R. Barro and X. Sala-i-Martin, Economic Growth (McGraw-Hill 1995), from which ao. the output growth rates λ and the capital's shares β have been taken for the period 1947-73, not included in the Penn Table. The numbers from their tables are marked BARRO.

– A. Maddison, Phases of Capitalist Development (OUP 1982) – the numbers taken from his tables are marked MADDISON.

– D.W. Jorgenson, F.M. Gollop and B. Fraumeni, Productivity and U.S. Economic Growth (North-Holland 1987). The numbers taken from their tables are marked JORGENSON.

No economic statistics gives estimates of the net savings rate s^*, because of the usually unknown depreciation. Here we shall use the following formula for the calculation of the average net savings rate of a certain period from the average gross savings rate S, which can be computed from the Penn tables:

$$s^* \approx \frac{10}{9}(S - .1),$$

It is based on the rough assumption recommended ao. by Samuelson and Nordhaus in their well-known textbook (Economics, the 1985 edition) that depreciation is about one tenth of the gross national product.

(Note that in our formulae Y of course is the net national product – hence the above formula.)

Another parameter, which cannot be derived from any economic statistics, is the discount rate ρ, which tells about how soon people want the chosen values come true. For the sake of simplicity, ρ is assumed to be .02 in all the cases studied in this book.

The parameter m must obey the condition (3.71). Therefore m must be chosen to be larger than $\alpha^*/(1-\beta)$ but very close to it.

Growth Effects of Savings Rate in the U.S.A

The BASIC PERIOD is 1947-73 and the PREDICTED PERIOD is 1960-73. Since the more comprehensive Penn tables start from 1950 the PENN numbers computed from the period 1950-73 are mostly used, except in the cases, where the BARRO numbers of the period 1947-73 are available, viz. in the cases of λ_1 and β.

Calibration (BASIC PERIOD 1947-73):

$\lambda_1 = \lambda_{1947-73}^{BARRO} = .0402$, $(\lambda - \dot{E}/E)_1 = (\lambda - \dot{E}/E)_{1950-73}^{PENN} = .018861 \Longrightarrow (\dot{E}/E)_1 = .021339.$

$\beta_{1947-73}^{BARRO} = .40$, $n_{1950-73}^{PENN} = .0136461$, $S_{1950-73}^{PENN} = .21929 \Longrightarrow s^* = .1325.$

$$\Longrightarrow b^* = .3033962, \quad \sigma = 3.8170814, \quad \beta/\sigma = .1047921$$

$$\Longrightarrow s^* - \beta/\sigma = .0277079, \quad \frac{\partial\alpha^*/\partial s^*}{\alpha^*} = -39.829107, \quad \alpha^* = .0811584$$

$$\Longrightarrow \frac{\alpha^*}{1-\beta} = .1352641. \Longrightarrow \text{Choose } m = .14.$$

Consider the contributions to λ_1 of capital, employment and TFP:

$$\lambda_1 = \beta\lambda_1 + (1-\beta)(\dot{E}/E) + TFP$$

$$\Longrightarrow .0402 = .01608 + .0128034 + TFP \Longrightarrow TFP = .0113168.$$

$$\Longrightarrow \kappa\nu = \kappa\left(\frac{1-\beta}{1-\beta+\kappa}\right)(\lambda_1 + m) = .0113168 \Longrightarrow \kappa = .0701422$$

$$\Longrightarrow \nu = .1613393$$

By using for the parameters in the equations (5)-(14) the values given by the above empirical estimates, we can now check the theory.

The Plosser fact:

We can now calculate:

$$1^{\circ}: \quad \frac{\beta b^*}{\sigma\alpha^*(s^* - \beta/\sigma)} = 14.138434$$

$$2^{\circ}: \quad \beta(\sigma - 1) - \alpha^* \left(\frac{\kappa}{1 - \beta + \kappa}\right)^2 (\lambda + m) = 1.1266723$$

Hence we get:

(6.19) $$\frac{\partial(Y^*/E^*)/\partial s^*}{Y^*/E^*} = 1^{\circ} \times 2^{\circ} = 15.929382 > 0.$$

Thus the Plosser effect is reproduced by the minimal macrodynamics constructed in Chapter 3 at least in the U.S. economy of the Golden Age period.

The quantitative growth effects of savings rate:

First we compute the following empirical estimate from the Penn Table:

$$(\lambda - \dot{E}/E)_2 = (\lambda - \dot{/E}/E)^{PENN}_{1960-73} = .0202159,$$

so that

(6.20) $$\Delta(\lambda - \dot{E}/E)_{obs} = .0013549.$$

On the other hand, by applying the parameter values obtained in the above calibration to (14), we get

(6.21) $$\frac{\partial Y^*/\partial s^*}{Y^*} = -23.943591,$$

after which (16) with the substitutions (19)-(21) gives the following theoretical prediction for the growth rate of output in the PREDICTION PERIOD :

(6.22) $$\lambda_{pred} = .038172, \quad \text{thus } \Delta\lambda_{pred} = \underline{-.002028}.$$

From the Penn Table we compute:

(6.23) $$\lambda_2 = \lambda^{PENN}_{1960-73} = .0375546, \quad \text{thus } \Delta\lambda_{obs} = \underline{-.0026454}.$$

In view of (22) and (23) the ratio (18) now gives:

Theory explains of observed change 77 %.

From (13),(17),(19) and (20) we get in a similar way:

$$\frac{\partial E^*/\partial s^*}{E^*} = -39.851029 \Longrightarrow \left(\frac{\dot{E}}{E}\right)_{pred} = .0179495,$$

so that

(6.24) $$\Delta(\dot{E}/E)_{pred} = \underline{-.0033895}.$$

A comparison with the empirical estimate computed from the Penn Table shows:

$$(\dot{E}/E)_2 = \lambda^{PENN}_{1960-73} - (\lambda - \dot{E}/E)^{PENN}_{1960-73} = .0173387, \text{ so that}$$

(6.25) $$\Delta(\dot{E}/E)_{obs} = \underline{-.0040003}.$$

In view of (24) and (25) the ratio (18)= gives:

Theory explains of observed change 85 %.

We can conclude that the empirics from the U.S. economy in the Golden Age period 1947-73 confirms rather well the extended growth effects of savings rate derived from the macroeconomic dynamics of Chapter 3, including the extraordinary twofold dependence on savings rate of output per worker, output and employment.

Growth Effects of Savings Rate in Japan

The BASIC PERIOD is again 1947-73 and PREDICTED PERIOD is 1960-73. To facilitate comparisons the calibration as applied here obeys the same rules as in the above case of the USA, the BARR0 numbers λ_1 and β however being now from the period 1952-73, which was in this case used in the Barro and Sala-i-Martin book (its Table 10.8).

Calibration (BASIC PERIOD 1950-73):

$\lambda_1 = \lambda^{BARRO}_{1952-73} = .0951$, $(\lambda - \dot{E}/E)^{PENN}_{1950-73} = .0565746$, $\Longrightarrow (\dot{E}/E)_1 =$ .0385254.

$\beta^{BARRO}_{1952-73} = .39,\ n^{PENN}_{1950-73} = .0106291,\ S^{PENN}_{1950-73} = .2795 \Longrightarrow s^* = .20.$

$\Longrightarrow b^* = .4755,\ \sigma = 1.9586035,\ \beta/\sigma = .1991214$

$\Longrightarrow s^* - \beta/\sigma = .0008786,\ \dfrac{\partial\alpha^*/\partial s^*}{\alpha^*} = -1143.4375,\ \alpha^* = .090345$

$\Longrightarrow \dfrac{\alpha^*}{1-\beta} = .1481065. \Longrightarrow$ Choose $m = .15.$

Consider the contributions to λ_1 of capital, employment and TFP:

$\lambda_1 = \beta\lambda_1 + (1-\beta)(\dot{E}/E)_1 + TFP$

$\Longrightarrow .0951 = .037089 + .0235004 + TFP \Longrightarrow TFP = .0345106$

$\Longrightarrow \kappa\nu = \kappa\left(\dfrac{1-\beta}{1-\beta+\kappa}\right)(\lambda+m) = .0345106 \Longrightarrow \kappa = .183055$

$\Longrightarrow \nu = .1885258.$

The Plosser fact:

We can now compute the values of the parameter functions

$$1^\circ:\ \frac{\beta b^*}{\sigma\alpha^*(s^* - \beta/\sigma)} = 1192.8155,$$
$$2^\circ:\ \beta(\sigma-1) - \alpha^*\left(\frac{\kappa}{1-\beta+\kappa}\right)^2(\lambda+m) = .3726756.$$

Hence we get

$$\frac{\partial(Y^*/E^*)/\partial s^*}{(Y^*/E^*)} = 1^\circ \times 2^\circ = 444.53323 > 0. \tag{6.26}$$

Thus the Plosser effect is there in the empirics obtained from the Japanese economy even stronger than in the U.S. economy, and well reproduced by the macroeconomic dynamics constructed in Chapter 3.

The quantitative growth effects of savings rate:

From the Penn Table we now get the empirical estimate

$(\lambda - \dot{E}/E)_2 = (\lambda - \dot{E}/E)^{PENN}_{1960-73} = .0659423,$

which gives

$$\Delta(\lambda - \dot{E}/E)_{obs} = .0093677. \tag{6.27}$$

The parameter values obtained in the calibration, when substituted in (14), now give

$$\frac{\partial Y^*/\partial s^*}{Y^*} = -707.74531, \tag{6.28}$$

after which (16) with the substitutions (26),(27) and (28) give the theoretical prediction for the growth rate of output in the PREDICTION PERIOD:

$$\lambda_{pred} = .0801666, \quad \text{thus } \Delta\lambda_{pred} = \underline{-.0149334.} \tag{6.29}$$

A comparison with the empirical estimate computed from the Penn tables, i.e.

$$\lambda_2 = \lambda_{1960-73}^{PENN} = .079486, \quad \text{thus } \Delta\lambda_{obs} = \underline{-.015614}, \tag{6.30}$$

shows that we now have the result:

Theory explains of observed change 96 %.

From (13),(17),(26) and (27) we get

$$\frac{\partial E^*/\partial s^*}{E^*} = -1146.6309 \Longrightarrow \left(\frac{\dot{E}}{E}\right)_2^{pred} = .014363, \tag{6.31}$$

so that

$$\Delta(\dot{E}/E)_{pred} = \underline{-.0241624}. \tag{6.32}$$

Again we compare this with the corresponding empirical estimate, obtained from the Penn Table:

$$(\dot{E}/E)_2 = \lambda_{1960-73}^{PENN} - (\lambda - \dot{E}/E)_{1960-73}^{PENN} = .0135437 \text{ so that}$$

$$\Delta(\dot{E}/E)_{obs} = \underline{-.0249817}. \tag{6.33}$$

It follows from (32) and (33) that the ratio (18) now again indicates that:

Theory explains of observed change 96 %.

We can conclude that the Japanese economy in the Golden Age period 1947-73 confirms still better than the U.S. economy the extended growth effects of savings rate, as derived from the macroeconomic dynamics of Chapter 3, and it thus confirms also again the extraordinary twofold dependence on savings rate of the output per worker Y/E, the output Y and the employment E. This dependence accordingly has already two mutually independent empirical confirmations.

Growth Effects of Savings Rate in Canada

This is the only example discussing the application of the theory to a relatively small nation – small in terms of population. Again the rule governing the calibration is based on the period 1947-73, with the BARRO numbers for λ_1 and β, the remaining empirical estimates being computed from the Penn tables.

Calibration (BASIC PERIOD 1947-73):

$\lambda_1 = \lambda_{1947-73}^{BARRO} = .0517$, $(\lambda - \dot{E}/E)_1 = (\lambda - \dot{E}/E)_{1950-73}^{PENN} = .021653 \Longrightarrow$ $(\dot{E}/E)_1 = .030047$

$\beta = \beta_{1947-73}^{BARRO} = .44$, $n = n_{1950-73}^{PENN} = .0193968$, $S = S_{1950-73}^{PENN} = .23471 \Longrightarrow s^* = .15$.

$$\Longrightarrow b^* = .3446666,\ \sigma = 4.0755497,\ \beta/\sigma = .1079608$$

$$\Longrightarrow s^* - \beta/\sigma = .0420392,\ \frac{\partial \alpha^*/\partial s^*}{\alpha^*} = -27.235597,\ \alpha^* = .0999533$$

$$\Longrightarrow \frac{\alpha^*}{1-\beta} = .178488 \Longrightarrow \text{Choose } m = .18.$$

Consider the contributions to λ_1 of capital, employment and TFP:

$$\lambda_1 = \beta\lambda_1 + (1-\beta)(\dot{E}/E)_1 + TFP$$

$$\Longrightarrow .0517 = .022748 + .0168263 + TFP \Longrightarrow TFP = .0121257$$

$$\Longrightarrow \kappa\nu = \kappa\left(\frac{1-\beta}{1-\beta+\kappa}\right)^2 (\lambda + m) = .0121257 \Longrightarrow \kappa = .0577277$$

$$\Longrightarrow \nu = .2100498.$$

The Plosser fact:

We get now the parameter functions:

$$1^\circ: \frac{\beta b^*}{\sigma\alpha^*(s^* - \beta/\sigma)} = 8.8555637,$$

$$2^\circ: \beta(\sigma - 1) - \alpha^* \left(\frac{\kappa}{1 - \beta + \kappa}\right)^2 (\lambda + m) = 1.3530395.$$

Hence:

$$\frac{\partial(Y^*/E^*)/\partial s^*}{(Y^*/E^*)} = 1^\circ \times 2^\circ = 11.576114 > 0. \tag{6.34}$$

Thus again, as in the cases of the U.S. and Japanese economies, the Plosser effect exists and is correctly reproduced by the theory.

The quantitative growth effects of savings rate:

The empirical estimate of the Canadian productivity of labour in the period 1960-73, as calculated from the Penn tables, is

$$(\lambda - \dot{E}/E)_2 = (\lambda - \dot{E}/E)^{PENN}_{1960-73} = .0241861.$$

which gives

$$\Delta(\lambda - \dot{E}/E)_{obs} = .0025331. \tag{6.35}$$

With the parameters given by the calibration we now get from (14):

$$\frac{\partial Y^*/\partial s^*}{Y^*} = -15.311198 \implies \lambda_2^{pred} = .0483499, \tag{6.36}$$

so that

$$\Delta\lambda_{pred} = \underline{-.0033501.} \tag{6.37}$$

As to the corresponding empirical value of λ_2, the estimates obtained from the Maddison and Penn tables differ rather much from each other here. They give the numbers

$$\lambda^{MADDISON}_{1960-73} = .0484479 \text{ and } \lambda^{PENN}_{1960-73} = .0493804,$$

respectively. Following the usual statistical trick we go the safest way and take the average:

$$\lambda_2 = .0489141 \text{ thus } \Delta\lambda_{obs} = \underline{-.0027858.} \tag{6.38}$$

It follows from (37) and (38) that the ratio (18) is now as large as 120.2 %. Thus we have the result:

Observed change is 83 % of predicted change.

Here of course the number 80 is the inverse of 120.

Let us study how the theory predicts the change in the growth rate of employment. With the substitutions (34),(36) and (37) the equation (13) gives:

$$(6.39)\qquad \frac{\partial E^*/\partial s^*}{E^*} = -27.25971 \implies (\dot{E}/E)_2^{pred} = .0240825,$$

so that

$$(6.40)\qquad \Delta(\dot{E}/E)_{pred} = \underline{-.0059645.}$$

By using the Maddison and Penn estimates for $\lambda_{1960-73}$ we get two different estimates of $\dot{E}/E)_{1960-73}$, viz. .0242618 and .0251943, respectively. Choosing again the average we have:

$$(6.41)\qquad (\dot{E}/E)_2 = .024728 \text{ thus } \Delta(\dot{E}/E)_{obs} = \underline{-.0053189.}$$

From (40) and (41) we now get the following result:

Observed change is 89 % of predicted change.

In other words, the theory a little overpredicts the observed change.

Comparing the rates of coverage 83 % and 89 % obtained here with the rates of explanation 77 % and 85 % obtained in the case of the U.S. economy we can say that the theory predicts in these two cases equally well, but not quite as well as in the case of Japanese economy. We can conclude however that the Canadian economy in the Golden Age period 1947-73 gives the third independent testimony to the existence of the extended growth effects of savings rate, as derived from the macroeconomic dynamics of Chapter 3, including the extraordinary twofold dependence on savings rate of the output per worker Y/E, the output Y and the employment E. This dependence accordingly has now had three mutually independent empirical confirmations.

The so far performed verifications of the growth effects of savings rate have all of them concerned countries outside Europe. We shall now investigate these growth effects in an European country, if only to show their limitations in explaining *all* the observed changes in the growth rates of productivity, employment and output.

Growth Effects of Savings Rate in France

The national economies of European countries do not in general satisfy the theoretical requirements of welldoing markets, especially labour markets. France is not an exception to the rule, but it will nevertheless be discussed here as an European example, to see the problems. Again the calibration follows the rule: the empirical estimates of λ_1 and β are taken from the book of Barro and Sala-i-Martin, and the rest of the needed estimates have been computed from the Penn Table.

*Calibration (*BASIC PERIOD 1950-73):

$\lambda_1 = \lambda_{1950-73}^{BARRO} = .0542$, $(\lambda - \dot{E}/E)_1 = (\lambda - \dot{E}/E)_{1950-73}^{PENN} = .0387248, \Longrightarrow$ $(\dot{E}/E)_1 = .0154752$

$\beta = .40$, $n = n_{1950-73}^{PENN} = .0092192$, $S = S_{1950-73}^{PENN} = .25495833 \Longrightarrow s^* =$ $.1721758$

$$\Longrightarrow b^* = .3147943, \ \sigma = 2.3547313, \ \beta/\sigma = .1698707$$

$$\Longrightarrow s^* - \beta/\sigma = .0023051, \ \frac{\partial\alpha^*/\partial s^*}{\alpha^*} = -438.21, \ \alpha^* = .0717177$$

$$\Longrightarrow \frac{\alpha^*}{1-\beta} = .1195295 \Longrightarrow \text{Choose } m = .12.$$

Consider the contributions to λ_1 of capital, employment and TFP:

$\lambda_1 = \beta\lambda_1 + (1-\beta)(\dot{E}/E)_1 + TFP$:

$$\Longrightarrow 0542 = .02168 + .0092851 + TFP \Longrightarrow TFP = .0232349.$$

$$\Longrightarrow \kappa\nu = \kappa\left(\frac{\kappa}{1-\beta+\kappa}\right)^2 (\lambda + m) = .0232349 \Longrightarrow \kappa = .1715062$$

$$\Longrightarrow \nu = .1354755.$$

The Plosser fact:

We get now the parameter functions:

$$1^\circ : \frac{\beta b^*}{\sigma\alpha^*(s^* - \beta/\sigma)} = 323.46649,$$

$$2^\circ : \beta(\sigma - 1) - \alpha^*\left(\frac{\kappa}{1-\beta+\kappa}\right)^2 (\lambda + m) = .5412751.$$

Hence:

$$\frac{\partial(Y^*/E^*)/\partial s^*}{(Y^*/E^*)} = 1^\circ \times 2^\circ = 175.08435 > 0.$$

Thus in all our four countries the Plosser effect of savings rate on the growth rate of output per worker is clearly reproduced by theory.

The quantitative growth effects of savings rate:

The empirical estimate of the French producivity of labour in the period 1960-73, as computed from the Penn tables, is

$(\lambda - \dot{E}/E)_2 = (\lambda - \dot{E}/E)^{PENN}_{1960-73} = .0401128.$

This gives

(6.42) $$\Delta(\lambda - \dot{E}/E)_{obs} = .001388.$$

With (43) and the parameters given by the calibration we get from (14):

(6.43) $$\frac{\partial Y^*/\partial s^*}{Y^*} = -264.18334 \implies \lambda_2^{pred} = .0521057,$$

so that

(6.44) $$\Delta\lambda_{pred} = \underline{-.0020943.}$$

The corresponding empirical estimates, computed from the Penn tables, are

(6.45) $$\lambda^{PENN}_{1960-73} = .0491753 \implies \Delta\lambda_{obs} = \underline{-.0050247.}$$

Because of (45) and (46) the ratio (18) now tells:

Theory explains of observed change only 42 %.

By substituting (43) and the parameter values from the calibration in (13) we get:

(6.46) $$\frac{\partial E^*/\partial s^*}{E^*} = -438.48919 \implies (\dot{E}/E)_2^{pred} = .011999,$$

so that

(6.47) $$\Delta(\dot{E}/E)_{pred} = \underline{-.0034762.}$$

to be compared with the corresponding empirical estimate computed from the Penn tables:

$$(6.48)\quad (\dot{E}/E)^{PENN}_{1960-73} = .0092192 \Longrightarrow \Delta(\dot{E}/E)_{obs} = \underline{-.006256.}$$

With the values from (48) and (49) the ratio (18) now gives the result:

Theory explains of observed value only 56 %.

The two numbers in the boxes show that the savings rate effect alone is not sufficient to explain the estimated changes in French productivity, employment and output: other factors too must have a remarkable part in the estimated changes. Some of them may be external influences on the French economy, but also the internal economic system plays a part: the European economies have traditionally been heavily regulated in a way where free markets are not necessarily guaranteed, especially in labour markets.

Conclusions

The verifications just performed testify

1) that the Plosser fact 7 is correctly reproduced by the canonical formalism of macroeconomics constructed in Chapter 3 in all the four countries tested above, and

2) that the extended effects of savings rate are quantified correctly by this theory, and verified by empirical statistics in the tested non-European capitalist countries USA, Japan, Canada. The only exception was the tested European country, France, where the quantitative effects of savings rate explained only half of the observed changes. Other external and internal factors are involved in this case.

The observed changes in the growth rates of output per worker, output and employment were rather small. When expressed in percentage units they were:

$\Delta(\lambda - \dot{E}/E)_{obs} = 0.14$ & (U.S.A.), 0.94 % (Japan), 0.25 % (Canada), 0.13 % (France),

$\Delta\lambda_{obs}$ = -0.26 % (U.S.A.), -1.56 % (Japan), -0.28 % (Canada), -0.50 % (France),

$\Delta(\dot{E}/E)_{obs}$ = -0.40 % (U.S.A.), -2.50 % (Japan), -0.53 % (Canada), -0.63 % (France).

Thus there is a third point in favour of the present theory, viz.

3) that the theory constructed in Chapter 3 functions exactly even in the case of small changes, in a sensitive and rather complicated mathematical environment.

The existence of the Plosser effect as well as the extended growth effects of savings rate are based on the twofold dependence on savings rate characteristic of the output per worker, output and employment in this theory. No such dependence is involved in any other growth theory. Indeed the twofold dependence on savings rate is an immediate consequence of the generalized value function underlying this theory. Thus we can conclude that the points 1-3 make of the growth effects of savings rate discussed in this chapter a decisive evidence for the macroeconomic dynamics constructed in Chapter 3.

Slowdown of Productivity and Savings Rate: A General Perspective

If we compare the growth rates of productivity and the level of savings rate in advanced countries with those countries, where economic growth is a relatively new phenomenon, we can observe a certain lawfulness. In advanced countries, where economic growth has been going on for a long time, the savings rate and the growth rate of productity of labour tend both of them to be small. The opposite is true in countries, whose economies have began to grow more recently, as shown in Table 2.

If we leave out Portugal and Turkey, where investments have not yet really began to grow, the difference between the two group of countries becomes still clearer: the newly growing countries have the average growth rate of productivitivity 4.1 % and the average net savings rate level .18.

The observed simultaneous slowdown of productivity and savings rate in the advanced countries with a long history of endogeneous economic growth of course is in harmony with the present theory. On the other hand, this is only a qualitative statement or a 'stylized fact', if you like. The slowdown itself may be a saturation phenomenon: in the

TABLE 2. – LONG-TERM TRENDS OF PRODUCTIVITY

Country:	Y/E-growth :	s^*-level :	Country	Y/E-growth :	s^*-level :
USA	1.4 %	.14	Japan	5.0 %	.255
GB	2.0 %	.09	Spain	3.9 %	.165
Canada	1.9 %	.145	Greece	4.1 %	.16
France	2.7 %	.18	Ireland	3.5 %	.155
Germany	2.5 %	.185	Portugal	4.1 %	.13
Holland	2.0 %	.155	Turkey	3.3 %	.12
Belgium	2.7 %	.14	–	–	–
Sweden	1.6 %	.14	–	–	–
Denmark	1.7 %	.165	–	–	–
Average:	1.9 %	.148	Average:	4.0 %	.165

advanced countries the infrastructure of economic growth is already there, which is why investments necessary to build it are no more needed. But the quantitative verification of the direct coupling between savings rate and the productivity of labour comes from the above calculations concerning the transitions from a BASIC PERIOD to a PREDICTED PERIOD .

There is also another case of simultaneous slowdown of productivity and savings rate in recent economic history of the advanced countries. As we know there followed, after the Golden Age period 1950-73, an irregular period 1974-82, during which both the productivity of labour and the savings rate fell in the advanced Western countries. In the USA, for instance, both the average growth rate of productivity and that of output fell from the period 1950-73 to the irregular period 1974-82, as computed from the Penn tables:

$$\begin{aligned}(\lambda - \dot{E}/E)^{PENN}_{1950-73} &= .0189 \longrightarrow (\lambda - \dot{E}/E)^{PENN}_{1974-82} = -.0026,\\ \lambda^{PENN}_{1950-73} &= .343 \longrightarrow \lambda^{PENN}_{1974-82} = .0160.\end{aligned}$$

This slowdown was also accompanied by a simulatenous decrease in the net savings rate, of the following order of magnitude:

$$s_{1950-73} \approx .13 \longrightarrow s_{1974-82} \approx .12.$$

Indeed, in a transition from any sufficiently long period to the period 1974-82 practically all Western economies show a negative change in the rate of growth of both the productivity and output, as well as a fall in the savings rate. However, all the observed slowdown in 1974-82 cannot be explained in terms of savings rate, there being the following general scheme of changes:

Change in the growth rate of	productivity	employment	output
Transition to the period 1974-82	−	+	−

It follows that external factors must have been involved in the observed changes of growth rates. In this case we indeed know what the main external factors were: the drastic rises of OPEC oil prices in 1973 and later in 1970s, called 'oil shocks'. Because of these remarkable external disturbances the observed changes in the growth rates of output and employment cannot be predicted from savings rates only.

7 The Kydland Facts on Leads and Lags in Business Cycles

(*The second decisive piece of evidence*)

7.1 The Importance of Labour Input and the Productivity of Labour

> "For growth, most of the output change is accounted for by changes in technology and in capital. In contrast, perhaps on the order of two-thirds of the business cycle is accounted for by movement in the labor input and one-third by changes in technology. Thus, most business cycle theorists agree that an understanding of aggregate labor market fluctuations is a prerequisite for understanding how business cycles propagate over time." (Finn Kydland)

The minimal macroeconomic dynamics constructed in Chapter 3 indeed suggests a solution to the problem of business cycles along those lines, emphasizing further the significance of the cycles in labor input and labor productivity. This concerns also the problem of leads and lags.

There are the facts 10-12 discussed by Kydland in the article from which the above quotation was taken[1]. As expressed by Kydland in the same article: "Average labor productivity is somewhat procyclical and leads the cycle ... The hours from the establishment survey indicate the longest lead: two to three quarters ... Total hours displays a slight phase shift in the direction of lagging the cycle, especially in the employment component."

More exactly the leads and lags can be studied from the cross-correlations, taken from Kydland's article and shown in Table 3. In this Table I have underlined the cross-correlations which are larger than the correlation for $t = 0$..

[1]The quotation is from 'Frontiers of Business Cycle Research', edited by Thomas F.Cooley and published by Princeton University Press 1995, p.126

TABLE 3. – CROSS-CORRELATIONS WITH OUTPUT: U.S. 1954-91.[2]

	$Y(-4)$	$Y(-3)$	$Y(-2)$	$Y(-1)$	Y	$Y(+1)$	$Y(+2)$	$Y(+3)$	$Y(+4)$
Y	.16	.38	.63	.85	1.00	.85	.63	.38	.16
Y/H	<u>.47</u>	<u>.51</u>	<u>.53</u>	<u>.44</u>	.32	−.06	−.30	−.47	−.50
E	−.03	.16	.38	.63	.83	<u>.88</u>	.80	.65	.46
H	−.07	.14	.39	.67	.88	<u>.91</u>	.80	.63	.42

Indeed, when contrasted with the beautiful symmetry of both sides in the case of the output Y in Table 3, the labour productivity here represented by Y/H shows a clear tendency to grow leftwards thus indicating a lead. The employment E surely shows a tendency to grow rightwards, thus indicating a lag but this lag is much smaller than the lead of labour productivity. Thus Table 3 mainly confirms the statements of Kydland. However it also checks and corrects the magnitudes of leads and lags involved:

(1) The labor productivity, for which we shall use the notation Y/E, leads the cycle, thus having its minimums and maximums before the output Y. The most probable lead is as large as 2-3 quarters of a year. But even the leads of 1 and 4 quarters are clearly possible.

(2) The employment E lags the cycle and thus has its minimums and maximums after the output Y. Reading the Table suggests that while the most probable lag is one quarter of a year, also a lag of one and a half quarters seems more probable than not.

(3) The 'Hours' H, i.e. the total working time of all people in the country, in the present theory represented by the variable $H = vN$, has its minimums and maximums close to those of the output Y and – perhaps – lags a little. But the difference between the one-quarter lag correlation .91 and the zero-lag correlation .88 is so small that it may be just a statistical error. All the leads and lags of course are statistical averages over all the observed cycles in the period 1954-91.

In the following it will be shown that the minimal macroeconomic dynamics constructed in Chapter 3 predicts correctly the lead of the

[2]The letter H refers here to the hours from the establishment survey, while the employment E is computed from the household survey; Y is the real gross national product, and the time unit is a quarter of a year.

labour productivity Y/E expressed in the statement (1) above. This is shown in three different periods in the U.S. economy.

The theory also predicts correctly, in all of those three periods, the lag of the employment E extending from one to one and a half quarters of a year, in agreement of the statement (2).

In the case of the hours H the theory will suggest that the movement of H around the cycle goes almost together with the output Y, which is not in disagreement with the facts expressed in the statement (3).

7.2 The Maximum and Minimum Points in Detrended Business Cycles

Imagine an observer who is situated at the fixed point P on the cycle plane (s, x), i.e. at the fixed center of ordinary business cycles (cf. Fig.2). Let us also imagine that the straight line $x = b^*$, which goes through the point P defines his horizon. The observer sees the state point (s, x) circle around him in its cyclic motion like a star in the sky. But he also can see the state $X(s, x)$ of each economic variable X circle around him doing its cycle. We want to know at which angle (with respect to the horizon) he sees the point at which any given economic variable X reaches its maximum (or minimum) in the course of its cycle. The world of this observer being two-dimensional such angles uniquely define the times at which each of those variables have their extremum values in the business cycle.

The extremum points of output, consumption and investment in the average business cycle:

Let us begin with the state (s, x) itself. We shall in the following of course use the second normal form, the equations (3.34), valid for the ordinary cycles around a balanced growth path, i.e.

$$-\dot{x}/x = (\beta - s)x - \alpha^*, \quad -\dot{s} = (\beta - \beta/\sigma)(1 - s)(x - b^*).$$

The extremum values of x $(=Y/K)$ are on the following curve defined by $\dot{x} = 0$:

$$0 = (\beta - s)x - \alpha^*.$$

This is a curve of the second order on the plane (s, x), in fact a hyperbola having the lines $x = 0$ and $s = \beta$ as its asymptotes (Fig.4).

Hence we get successively the function $x(s)$ and its derivative:

$$x = \frac{\alpha^*}{\beta - s}, \quad \frac{dx}{ds} = \frac{\alpha^*}{(\beta - s)^2} > 0.$$

Thus the angles, with respect to the straight line $x = b^*$, at which an observer at the cycle center sees the maximums and the minimums are as follows:

$$\text{(7.1)} \qquad \text{Max } x : \varphi x = \arctan\left[\frac{\alpha^*}{(\beta - s^*)^2}\right], \text{Min } x : \varphi x - 180^o.$$

Once we know the angle in which the extremum points of x are seen from the cycle center P, it is easy to transform this knowledge to units of time. The angle in which a whole cycle is seen at the cycle center P is of course 360°, which accordingly corresponds to the length of a cycle in time units. We shall choose an average length of a business cycle to be 4 years, which is not a bad approximation of reality. Thus an angle of 90° approximately corresponds to one year, and an angle $\arctan dx/ds$ corresponds to the fraction $(\arctan dx/ds)/90^o$ of a year.

Because the cycle functions of the output Y, the consumption C and, in a linear approximation valid in a neighbourhood of the cycle center P, also the cycle function of the investment I are all of them proportional to $x - b^*$, there being

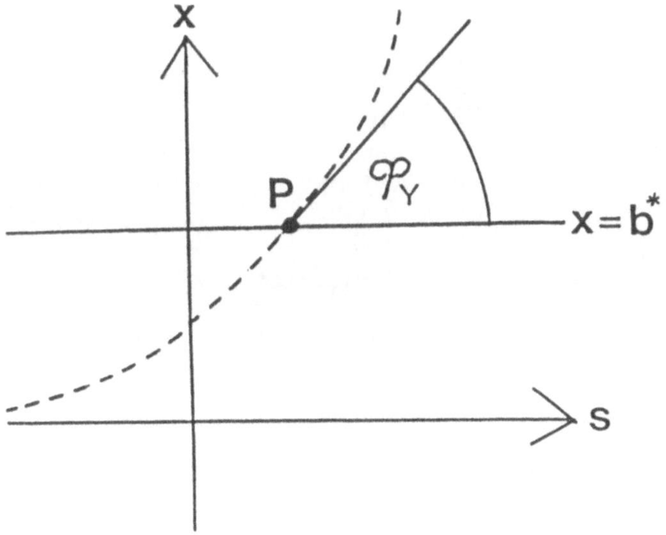

FIGURE 4. – THE CURVE DEFINED BY $\dot{x} = 0$ ON THE STATE-PLANE (s, x).

$$Q_Y = \beta(x - b^*),\ Q_C = Q_Y/\sigma_,,\ Q_I = \left[1 + \left(\tfrac{\sigma - 1}{\sigma}\right)\left(\tfrac{1 - s^*}{s^*}\right)\right] Q_Y,$$

they have their extremum points simultaneously with the output/capital ratio x.

The cycle function of investment is nonlinear, and if the first nonlinear terms were taken into account, there would appear a typical nonlinear effect in the maximum and minimum points: the maximum could show a lag while the minimum shows a lead, or vice versa. Here these nonlinear effects will be passed by.

The extremum points of employment in the average business cycle:

We have now to study the cycle function of employment,

$$Q_E = \left(\frac{\beta}{1 - \beta + \kappa}\right)[(1 - s)x - (1 - s^*)b^*].$$

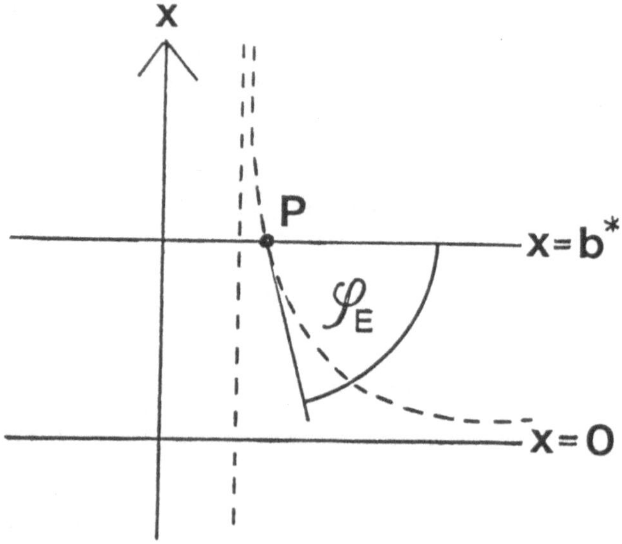

FIGURE 5. – THE CURVE DEFINED BY $\dot{Q}_E = 0$ ON THE PLANE (s, x).

From (3.34) we first derive the following formula:

$$\frac{d}{dt}(1-s)x = (1-s)x\,[(\beta-s^*)b^* - (\beta-s)x + (\beta-\beta/\sigma)(1-s)(x-b^*)\,].$$

Hence we get: $\dot{Q} = 0$ on the curve

$$(\beta-s^*)b^* - (\beta-s)x + (\beta-\beta/\sigma)(1-s)(x-b^*) = 0.$$

This is again a hyperbola. Fig.5 shows its course near the point P for an economy of the second half of century. On the first half the curve is similar but on other other side of the line $s = s^*$. By derivation with respect to s and substituting then $s = s^*, x = b^*$ we get:

$$\left(\frac{dx}{ds}\right)_P = -\frac{b^*}{(s^*-\beta/\sigma)-(\beta-\beta/\sigma)s^*}.$$

It follows that the extremum points of the employment cycle function Q_E are, when seen from the cycle center P, in the directions defined by the following angles:

$$(7.2)\,\varphi_E = -\arctan\left[\frac{b^*}{(s^*-\beta/\sigma)-(\beta-\beta/\sigma)s^*}\right] \text{ and } \varphi_E - 180^o,$$

where the positive extremum is the maximum point, and negative the minimum point.

The extremum points of labour productivity in the average business cycle:

The labour input is equal to E, which is why the productivity of labour is correctly represented by Y/E. Thus we have to study the cycle function

$$Q_{Y/E} = Q_Y - Q_E = \beta(x-b^*) - \left(\frac{\beta}{1-\beta+\kappa}\right)[(1-s)x-(1-s^*)b^*].$$

By derivation with respect to time and taking the derivatives $\dot{x}$ and $\dot{s}$ from the cycle equations (3.34) we get the extremum value condition $\dot{Q}_{Y/E} = 0$ in the following form:

$$\alpha^*-(\beta-s)x = \left(\frac{1}{1-\beta+\kappa}\right)(1-s)[(\beta-s^*)b^*-(\beta-s)x+(\beta-\beta/\sigma)(1-s)(x-b^*)].$$

This accordingly is the curve of the third order on which are the extremum points of $Q_{Y/E}$.

By derivation with respect to s and making after that the substitutions $s = s^*$ and $x = b^*$ we now get:

$$\left(\frac{dx}{ds}\right)_P = \frac{[1-\beta+\kappa]-(1-s^*)]b^*}{(1-\beta+\kappa)(\beta-s^*)+(1-s^*)(s^*-\beta/\sigma)-(1-s^*)s^*(\beta-\beta/\sigma)},$$

which may be positive or negative depending on the values of the parameters. It will be shown later that the maximum and minimum points of Y/E are as follows:

$$\text{(7.3)} \quad \text{Max } Q_{Y/E} : \varphi_{Y/E} =$$
$$= \arctan\left[\frac{[(1-\beta+\kappa)-(1-s^*)]b^*}{(1-\beta+\kappa)(\beta-s^*)+(1-s^*)(s^*-\beta/\sigma)-(1-s^*)s^*(\beta-\beta/\sigma)}\right],$$
$$\text{(7.4)} \quad \text{Min } Q_{Y/E} : \varphi_{Y/E} - 180^o.$$

The extremum points of total working hours
in the average business cycle:

The maximum and minimum points of the total working hours, or simply 'Hours' $H = vN = (\alpha^*/k) + (\dot{h}/h)/k$, are on the curve defined by the equation

$$\frac{dH}{dt} = \frac{1}{k}\left[\frac{d}{dt}(\dot{h}/h) - m(\dot{h}/h) - \alpha^* m\right] = 0.$$

Substituting here $\dot{h}/h = \nu + Q_h = \nu + Q_E$ we get, after some calculation:

$$(1-s)x\,[(\beta-s^*)b^*-(\beta-s)x+(\beta-\beta/\sigma)(1-s)(x-b^*)-m] = \text{Constant}.$$

This defines a curve of the fourth order on the plane (s, x).

The derivation with respect to s now gives:

$$\frac{dx}{ds}\left\{(1-s)\,[\cdot]-(1-s)(\beta-s)x+\left(\beta-\frac{\beta}{\sigma}\right)(1-s)^2x\right\}$$
$$= x\,[\cdot]-(1-s)x^2+\left(\beta-\frac{\beta}{\sigma}\right)(1-s)x(x-b^*).$$

Substituting here $s = s^*$ and $x = b^*$ we get the result:

$$\left(\frac{dx}{ds}\right)_P = \frac{mb^* + (1-s^*)(b^*)^2}{(1-s^*)m - (1-s^*)\left[(s^* - \beta/\sigma) - (\beta - \beta/\sigma)s^*\right] b^*}$$

The angles indicating the direction of maximum and minimum points of total working hours will be:

(7.5) Max $H : \varphi_H =$

$$\arctan\left\{\frac{mb^* + (1-s^*)(b^*)^2}{(1-s^*)m - (1-s^*)\left[(s^* - \beta/\sigma) - (\beta - \beta/\sigma)s^*\right] b^*}\right\},$$

(7.6) Min $H : \varphi_H - 180^o$.

7.3 The Leads and Lags in the Second Half of the Century

After those calculations we are ready to test the theory by comparing the predicted leads and lags with what is known empirically of the leads and lags in the business cycle. We can show that the stylized facts known on the leads and lags in the business cycle are very well predicted by the present theory. Indeed to my knowledge this theory is the only one capable of their quantitative explanation.

We shall try the theory by using different calibrations, in order to see how much the predictions depend on the chosen values of parameters. The empirical stylized facts as reported before refer to the business cycles observed during the period after the second world war. We shall use two calibrations belonging to that period and, to be sure of the predictions, later even one that refers mainly to the first half of this century.

The two calibrations from the second half of the century are performed using the periods 1947-73 and 1950-81. We begin with the Golden Age ending in 1973, and consider after that what happens to the leads and lags when the irregular years 1974-81 are added to the period of calibration.

The stylized facts concern the U.S. economy, which also has been the essential target of the modern theories of the business cycle. Thus the calibrations given here also come from the U.S. economy.

Leads and Lags in the Golden Age period 1947-73

In conformity with Chapter 6 the calibration applies the values of the growth rate of output λ and the capital's share β given by Barro and Sala-i-Martin (1995), while the other values of parameters are those computed from the Penn tables (the net savings rate, as before, by using the formula $s^* \approx (10/9)(S - .1)$ where S is the computed gross savings rate).

Calibration:

$$\lambda = .0402, \quad \beta = .40, \quad s^* = .1325 \Longrightarrow b^* = .3033962, \quad \sigma = 3.8170814,$$

$\Longrightarrow s^* - \beta/\sigma = .0277079, \quad \alpha^*/(1-\beta) = .135 \Longrightarrow m = .14, \quad \kappa = .0701422.$

The predicted maximum and minimum points:

With the above values of parameters we get the following angles φ_X at which the maximum and minimum points of each variable X is seen from the cycle center P:

$$\underline{\varphi_Y = \varphi_C = \varphi_I = \varphi_{Y/K} =}$$

$$= \arctan\left[\frac{b^*}{\beta - s^*}\right] = \arctan(1.1341914) = 48.6^o \ : \ \text{Maximum,} \tag{7.7}$$

$$\varphi_Y - 180^o = -131.4^o \ : \ \text{Minimum,} \tag{7.8}$$

$$\underline{\varphi_E} = -\arctan\left[\frac{b^*}{(s^* - \beta/\sigma) - (\beta - \beta/\sigma)s^*}\right] =$$

$$= \arctan(26.597) = 87.85^o \ : \ \text{Maximum,} \tag{7.9}$$

$$\varphi_E - 180^o = -92.15^o \ : \ \text{Minimum,} \tag{7.10}$$

$$\underline{\varphi_{Y/E}} =$$

$$= \arctan\left[\frac{[(1-\beta+\kappa)-(1-s^*)]b^*}{(1-\beta+\kappa)(\beta-s^*)+(1-s^*)(s^*-\beta/\sigma)-(1-s^*)s^*(\beta-\beta/\sigma)}\right] =$$

$$= -\arctan(.3535369) = -19.5^o \ : \ \text{Maximum,} \tag{7.11}$$

$$\varphi_{Y/E} + 180^o = 160.5^o \ : \ \text{Minimum} \tag{7.12}$$

$$\underline{\varphi_H} =$$

$$\arctan\left[\frac{\mathrm{m}b^* - (1-\mathrm{s}^*)(\mathrm{b}^*)^2}{(1-\mathrm{s}^*)\mathrm{m} - (1-\mathrm{s}^*)[(\mathrm{s}^* - \beta/\sigma) - (\beta - \beta/\sigma)\mathrm{s}^*]\mathrm{b}^*}\right] =$$

$$= \arctan(.9829308) = 44.5^o \ : \ \text{Maximum,} \tag{7.13}$$

$$\varphi_H - 180^o = -135.5^o \ : \ \text{Minimum.} \tag{7.14}$$

Before proceeding to comparisons of these theoretical predictions with empirical facts, a possible puzzle in the above Maximum/Minimum numbers perhaps must be discussed. The reader may wonder: why do we know that $\varphi_{Y/E} = -19.5^o$ indicates a maximum point and not a minimum point? This obviously is the only case, where a question about the above numbers may arise. However, it follows from the relation

$$\dot{Q}_{Y/E} = \dot{Q}_Y - \dot{Q}_E$$

that for $\dot{Q}_Y = 0$ we have $\dot{Q}_{Y/E} = -\dot{Q}_E$. In other words: when Q_Y reaches its maximum at 48.6^o, the other two are developing in opposite directions. But we know that at the angle $\varphi_Y = 48.6^o$ the employment E still grows toward its Maximum at 87.85^o. It follows that the productivity of labour Y/E is decreasing from its extreme value at -19.5^o, which accordingly must be its maximum point.

Comparison with the empirical facts:

The theoretical predictions just computed are illustrated in Fig.6. As mentioned already in Chapter 3 the state point (s, x) in the ordinary growth cycles revolves counterclockwise, as indicated in the picture. This indeed is the case, which determines the true direction of revolution in the ordinary business cycles: only a counterclockwise direction matches the empirical facts, and matches them perfectly.

The Kydland facts concerning the leads and lags in the business cycle were given in Section 7.1.

We can see from Fig.6 that the productivity of labour Y/E leads the output Y. But how much? The question can be answered as follows. As we know the periods of the business cycles vary greatly. However, the average length of the business cycle in this century is usually estimated to be close to 4 years. Accepting this average a total cycle in Fig.6 corresponds to 4 years and accordingly an angle of 90° corresponds approximately to a year. It follows that the lead of productivity with respect to output, viz. $48.6^o + 19.5^o = 68.1^o$ corresponds to $68/90(= 0.75)$ years, which is equal to 9.1 months, or 3.0 quarters. This is on the border of the nuclear range of two to three quarters of a year and we have here an *excellent* prediction success.

The theory also predicts a lag of the employment E. By using the same method as above we can estimate its length in time by computing first its length as an angle, which is equal to $87.85^o - 48.6^o = 39.25^o$. This corresponds to $39.25/90 = 0.44$ years, i.e. 5.3 months or 1.75

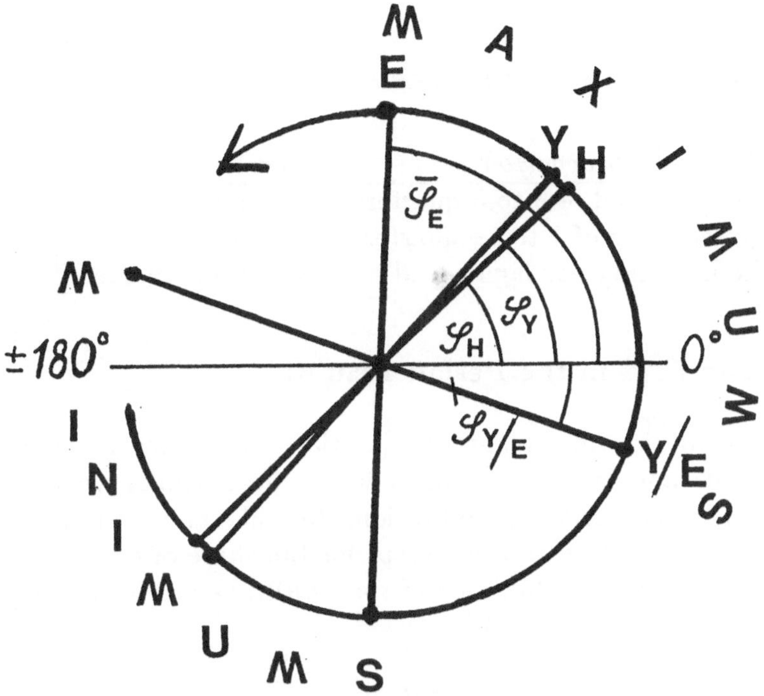

FIGURE 6. – THE LEADS AND LAGS PREDICTED BY THE THEORY FOR THE GOLDEN AGE PERIOD 1947-73.

quarters. This exceeds the empirically observed region of 1 to $1\frac{1}{2}$ quarters but only by 0.25 quarters, i.e. about three weeks. This surely is within the error of measurement, and we must consider the predicted value as *good* even though not excellent. The predicted lag of employment is clearly smaller than the predicted lead of productivity, in conformity with empirical observations.

The third stylized fact concerning the cycle of the total hours H as being very close to the cycle of output itself, is rather well predicted theoretically, which suggests a lead as small as $48.6^o - 44.5^o = 4.1^o$ corresponding to 2.3 weeks or 0.2 quarters. This certainly is within the statistical error of the stylized fact "no clear lag no clear lead", and we can consider it as *good*, though not as excellent.

To sum up, we have the results indicated in Table 4 concerning the empirical observations and the corresponding theoretical predictions pertaining to the business cycle in the Golden Age period.

TABLE 4. – LEADS AND LAGS IN THE GOLDEN AGE PERIOD.

Variable :	*Observation :*	*Prediction :*	*Appraisal :*
Productivity	lead of 2 to 3 quarters	lead of 3 quarters	excellent
Employment	lag of 1 to $1\frac{1}{2}$ quarters	lag of 1.75 quarters	good
Total hours	no clear lag or lead	lead of 0.2 quarters	good

Leads and Lags in the Period 1950-81

This analysis will answer the question about how much the predicted leads and lags are changed when the irregular years 1974-81 are added to the period of calibration. Just like above the parameters are based on the Penn tables, except for the share of capital β, whose value is a weighted combination of the BARRO value (1947-73) and the JORGENSEN value (1974-81).

Calibration:

$\lambda = .0304678$, $\beta = (27/33)\beta^{BARRO}_{1947-73} + (6/33)\beta^{JORGENSON}_{1974-81} = .3967$, $s^* = .1302 \Longrightarrow b^* = .2340076$, $\sigma = 4.10504$,

$\Longrightarrow s^* - \beta/\sigma = .0335627$, $\alpha^*/(1-\beta) = .10337 \Longrightarrow m = .1034$, $\kappa = .0688928$.

The predicted maximum and minimum points:

By using the above values of the parameters we get the following angles φ_X at which the maximum and minimum points of each variable X are seen from the cycle center P:

$\underline{\varphi_Y = \varphi_C = \varphi_I = \varphi_{Y/K} =}$

(7.15) $\arctan\left[\dfrac{b^*}{\beta - s^*}\right] = \arctan(.8780772) = 41.3^o$: Maximum,

(7.16) $\varphi_Y - 180^o = -138.7^o$: Minimum,

$\underline{\varphi_E} = -\arctan\left[\dfrac{b^*}{(s^* - \beta/\sigma) - (\beta - \beta/\sigma)s^*}\right] =$

(7.17) $+\arctan(42.504613) = 88.65^o$: Maximum,

(7.18) $\varphi_E - 180^o = -91.35^o$: Minimum,

$\underline{\varphi_{Y/E}} =$

$\arctan\left[\dfrac{[(1-\beta+\kappa)-(1-s^*)]b^*}{(1-\beta+\kappa)(\beta-s^*)+(1-s^*)(s^*-\beta/\sigma)-(1-s^*)s^*(\beta-\beta/\sigma)}\right] =$

(7.19) $-\arctan(.2652211) = -14.85^o$: Maximum,

(7.20) $\varphi_{Y/E} + 180^o = 165.15^o$: Minimum,

$\underline{\varphi_H} =$

$\arctan\left[\dfrac{mb^* - (1-s^*)(b^*)^2}{(1-s^*)m - (1-s^*)[(s^*-\beta/\sigma)-(\beta-\beta/\sigma)s^*]b^*}\right] =$

(7.21) $\arctan(.7887961) = 38.3^o$: Maximum,

(7.22) $\varphi_H - 180^o = -141.7^o$: Minimum.

Comparison with empirical facts:

The theoretical predictions just computed are illustrated in Fig.7.The configuration as a whole is very similar to that in Fig.6. We can again see the lead of the productivity of labour Y/E and the lag of the employment E, in accordance with the facts. The predicted lead of productivity is $41.3^o + 14.85^o = 56.15^o$, equal to 0.62 years, i.e. 2.5 quarters. This is within the nuclear region, which is two or three quarters, or 6 to 9 months.

The predicted lag of employment is now as large as $88.65^o - 41.3^o = 47.35^o$ or 0.52 years, i.e. 2.1 quarters, which is too much. No success of prediction here. But the total hours again go rather close the output in their theoretical cycle, the difference between them being only $41.3^o - 38.3^o = 3.0^o$. This is about two weeks, or 0.1 quarters, which certainly is within the range of statistical error of the observed "no clear lag no clear lead".

These numbers, as well as Fig.7, testify that the irregular years 1974-81 did not change much the leads and lags, except for increasing the lag of employment, as indicated in Table 5.

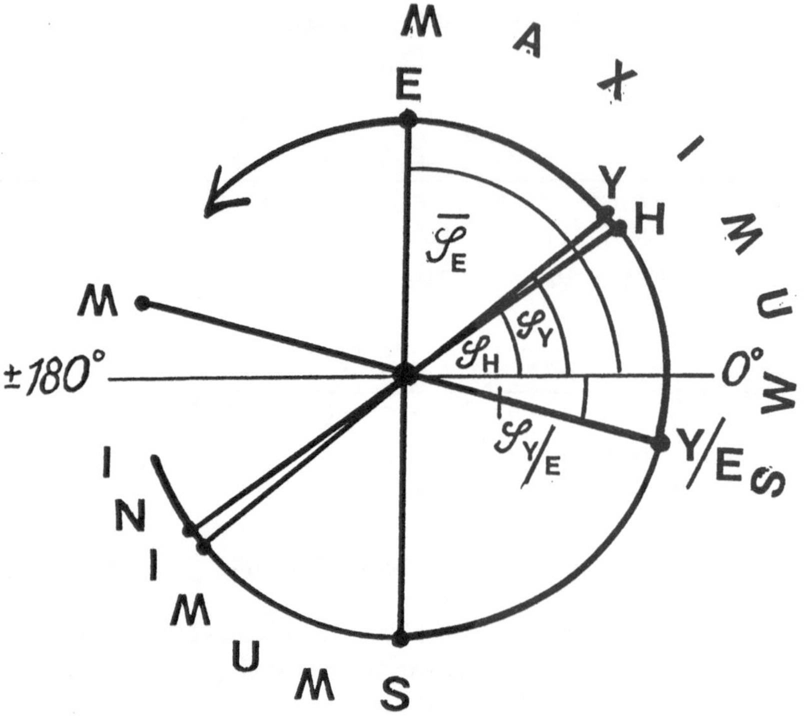

FIGURE 7. – THE LEADS AND LAGS PREDICTED BY THE THEORY FOR THE PERIOD 1950-81.

TABLE 5. – LEADS AND LAGS IN THE PERIOD 1950-81.

Variable :	*Observation* :	*Prediction*	*Appraisal* :
Productivity	lead of 2 to 3 quarters	lead of 2.5 quarters	excellent
Employment	lag of 1 to $1\frac{1}{2}$ quarters	lag of 2.1 quarters	bad
Total hours	no clear lag or lead	lead of 0.1 quarters	good

7.4 The Leads and Lags in the First Half of the Century

How much the business cycle has changed during the course of this century? To get some answer to this question a mixed parametrization will be used, where three of the basic five parameters refer directly to the first half of this century. These parameters are: 1) the traditional share of capital $\beta = 1/4$ coming from the classic Denison study (1961) concerning the period 1909-57, 2) the average output/capital ratio $b^* = .30$ taken from Solow's 1957 paper and valid from 1909 to 1930 as illustrated in Fig.3, and 3) the growth rate of population $n \approx .015$. Furthermore, a fourth parameter κ has been given the value .417, which differs widely from the calibrations in this book based on the Penn tables but was chosen by Lucas in his seminal 1988 paper (κ is the parameter that Lucas called γ).

Calibration:

$$\beta = .25,\ \ b^* = .30\ \ s^* = .13\ \ n = .015,\ \ \kappa = .417$$

$$\Longrightarrow \lambda = .039,\ \ \sigma = 2.2917,\ \ s^* - \beta/\sigma = .0209106,$$

$$\alpha^*/(1-\beta) = .048 \Longrightarrow m = .05.$$

The predicted maximum and minimum points:

The above values of the parameters give now the following maximum and minimum points of the economic variables in the business cycle:

$$\underline{\varphi_Y = \varphi_C = \varphi_I = \varphi_{Y/K} =}$$

$$\text{(7.23)}\quad \arctan\left[\frac{b^*}{\beta - s^*}\right] = \arctan(2.5) = 68.2^o\ :\ \text{Maximum},$$

$$\text{(7.24)}\quad \varphi_Y - 180^o = -111.8^o\ :\ \text{Minimum},$$

$$\underline{\varphi_E} =$$

$$-\arctan\left[\frac{b^*}{(s^* - \beta/\sigma) - (\beta - \beta/\sigma)s^*}\right] = -\arctan(115.72734)$$

$$\text{(7.25)}\quad = -89.5^o\ :\ \text{Minimum},$$

$$\text{(7.26)}\quad \varphi_E + 180^o = 90.5^o\ :\ \text{Maximum},$$

$$\underline{\varphi_{Y/E}} =$$

$$\arctan\left[\frac{[(1-\beta+\kappa)-(1-s^*)]b^*}{(1-\beta+\kappa)(\beta-s^*)+(1-s^*)(s^*-\beta/\sigma)-(1-s^*)s^*(\beta-\beta/\sigma)}\right]$$

$$= \arctan(.626163) = 32.0^o \ : \ \text{Maximum},$$

$$\varphi_E - 180^o = -148.0^o \ : \ \text{Minimum},$$

$$\underline{\varphi_H} =$$

$$\arctan\left[\frac{mb^* - (1-s^*)(b^*)^2}{[(1-s^*)m - (1-s^*)[(s^*-\beta/\sigma)-(\beta-\beta/\sigma)s^*]b^*}\right]$$

(7.27) $$= \arctan(2.1787153) = 65.3^o \ : \ \text{Maximum},$$

(7.28) $$\varphi_H - 180^o = -114.7^o \ : \ \text{Minimum}.$$

Comparison with the empirical facts:

The above theoretical predictions obtained from the calibration with emphasis on the first half of the century are shown in Fig.8. Comparing it with Figs. 6 and 7 we can see that now the configuration as a whole has travelled a little counterclockwise, i.e. in the direction of the cycle motion. This time the lead of productivity is only 36.2^o which corresponds to 0.4 years, i.e. 5 months or 1.6 quarters. This misses the nucleus of observed leads, 2 to 3 quarters, just by a month, but is well within the possible values from 1 to 4 quarters. Accordingly we can record it as *good.*

The lag of employment is predicted to be 22.3^o which is equal to 0.25 years or precisely one quarter. This is in full agreement with the empirical observation. Thus we have here an *excellent* prediction success.

The theory again predicts, for the third time already, a small lead of the total hours. This lead is only 2.9^o which corresponds to 0.4 months or 0.1 quarters. The observed fact "no clear lag no clear lead" most probably has a statistical error of at least one month, and the present result can still be called a *good* representation of the observed fact.

Table 6 and Fig.8 show that the configuration of leads and lags was in the first half of the century only $1.6 + 1 = 2.6$ quarters long, which is much less than the $2.1 + 2.5 = 4.6$ quarters in the period 1950-81 (Table 5) or the $1.75 + 3 = 4.75$ quarters in the Golden Age period (Table 4).

This is what one could expect, since the theory suggests a period of anomalous cycles to be associated with the crash of 1929, *shortening the leads and lags in the first half of the century.*

TABLE 6. – LEADS AND LAGS IN THE FIRST HALF OF THE CENTURY

Variable :	*Observation* :	*Prediction* :	*Appraisal* :
Productivity	lead of 2 to 3 quarters	lead of 1.6 quarters	good
Employment	lag of 1 to $1\frac{1}{2}$ quarters	lag of one quarter	excellent
Total hours	no clear lag or lead	lead of 0.1 quarters	good

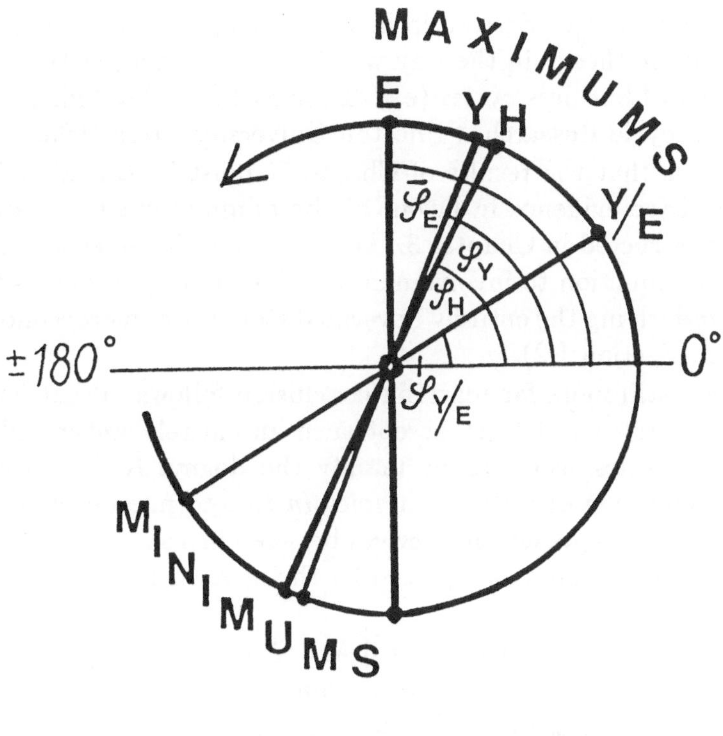

FIGURE 8. – THE LEADS AND LAGS PREDICTED BY THE THEORY FOR THE FIRST HALF OF THE CENTURY.

Conclusions from the Analysis of Leads and Lags

Eight out of nine empirical verifications gave results that must be considered as excellent or good representations of the Kydland facts. This is a prediction success, which is far above anything reached by any other theory of the business cycle today.

Hence we come to a major conclusion from the above analysis of leads and lags. This successful analysis was based on the form of the cycle functions $Q_X(s,x)$. Obviously the resulting maximum and minimum points of the variables X, and thus the leads and lags of these variables with respect to the output Y, reflect the exact mathematical properties of the respective cycle functions, which were derived from the generalized value function of macroeconomics introduced in Chapter 3. It is difficult to imagine that the same cycle functions could be derived from any theory that differs essentially from the present one. Surely they cannot be derived from stochastic processes superposed on microeconomic theory in the way which today dominates the theoretical research of business cycles (e.g. Thomas F. Cooley (ed.), Frontiers of Business cycle Research, Princeton University Press 1995).

It follows that the results of Chapter 7 must be considered as the second decisive evidence in favour of the minimal macroeconomic dynamics constructed in Chapter 3. This dynamics includes an extension of the value function to involve nonmaterial values but rejects the two dogmas underlying the entirely unverified stochastic-microeconomic approach (cf. Section 1.2).

Hence a still more far-reaching conclusion follows. What has made the economists so widely to accept such an entirely unverified theory of business cycles seems to be mainly the dogma No. 1, which demands a reduction of macroeconomics *in toto* to microeconomics. i.e. to the exchange of goods and services between firms and households. In this book this pursuit was replaced by the introduction of nonmaterial values.

The results of both Chapters 6 and 7 suggest that the economic relevance of nonmaterial values *in the long-term economic development*, including the business cycles, has been now sufficiently proved and that the relevance of the stochastic processes superposed on microeconomic theory is limited to short-term development.

8 Correlations and Variances over the Ordinary Business Cycles

(*Complementary Evidence*)

8.1 Calculation of Theoretical Correlations and Standard Deviations

The choice between the two normal forms of cycle equations is irrelevant in Chapter 8 (because of their symmetry) – we shall here use the first of them, (3.29), since it applies in Chapter 9 too.The solution of the linear approximation (3.46)-(3.47) can be written in the form

$$(8.1)\quad s - s^* \stackrel{Lin}{=} C\,e^{ut} + \bar{C}\,e^{\bar{u}t}, \quad C = \left(\frac{1}{2} + \frac{i\alpha}{4\omega}\right)[s(0) - s^*],$$

$$(8.2)\quad x - b^* \stackrel{Lin}{=} D\,e^{ut} + \bar{D}\,e^{\bar{u}t}, \quad D = \left[\frac{ia(\alpha^2 + 4\omega^2)}{8(1 - s^*)\omega}\right][s(0) - s^*].$$

Here $u = \alpha/2 + i\omega$, $\bar{u} = \alpha/2 - i\omega$, and $\alpha = \alpha^*$, there being

$$(8.3)\quad \alpha = (\beta - s^*)b^*, \quad \omega = b^*\sqrt{\left(\frac{1 - s^*}{a}\right) - \left(\frac{\alpha}{2b^*}\right)^2}, \quad a = \frac{\sigma}{\beta(\sigma - 1)}.$$

Direct calculation gives the 'basis integrals':

$$I_S = \int (s - s^*) = \left(\frac{\alpha}{\omega^2 + \alpha^2/4}\right) \cdot \left(e^{\alpha\pi/\omega} - 1\right)[s(0) - s^*],$$

$$I_x = \int (x - b^*) = \left(\frac{a}{1 - s^*}\right) \cdot \left(e^{\alpha\pi/\omega} - 1\right)[s(0) - s^*],$$

$$I_{SS} = \int (s - s^*)^2 = F_{SS}(\alpha, \omega) \cdot \left(e^{2\alpha\pi/\omega} - 1\right)[s(0) - s^*]^2,$$

$$(8.4)\quad F_{SS} = \left[\frac{\alpha^2 + 4\omega^2}{8\alpha\omega^2} + \alpha\left(\frac{\frac{3}{2} - \alpha^2/8\omega^2}{\alpha^2 + 4\omega^2}\right)\right],$$

$$I_{Sx} = \int (s - s^*)(x - b^*) = F_{Sx}(a, s^*) \cdot \left(e^{2\alpha\pi/\omega} - 1\right)[s(0) - s^*]^2,$$

$$(8.5)\quad F_{Sx} = \left[\frac{a}{2(1 - s^*)}\right],$$

$$I_{xx} = \int (x - b^*)^2 = F_{xx}(a, \alpha, \omega, s^*) \cdot \left(e^{2\alpha\pi/\omega} - 1\right) [s(0) - s^*]^2,$$

$$F_{xx} = \frac{a^2(\alpha^2 + 4\omega^2)^2}{32(1 - s^*)^2\omega^2} \left(\frac{1}{\alpha} - \frac{\alpha}{\alpha^2 + 4\omega^2}\right). \tag{8.6}$$

All the integrals are over the cycle, i.e. from $t = 0$ to $t = T = 2\pi/\omega$.

A linear approximation will be used also for the cycle functions, where now $b = b^*$. By developing each of them in terms of the powers of $s - s^*$ and $w - b^*$ and accepting only the linear terms we get:

$$Q_X^* \overset{Lin}{=} L_X(s - s^*, x - b^*) = A_X(s - s^*) + B_X(x - b^*)$$

for the cycle function Q_X^* of each variable X. Here

$$A_Y = 0, \quad B_Y = \beta, \quad A_C = 0, \quad B_C = \beta/\sigma, \tag{8.7}$$

$$A_I = 0,\ B_I = \beta\left[1 + \left(\frac{\sigma - 1}{\sigma}\right)\left(\frac{1 - s^*}{s^*}\right)\right],\ A_E = -\left(\frac{\beta}{1 - \beta + \kappa}\right) b^*; \tag{8.8}$$

$$B_E = \left(\frac{\beta}{1 - \beta + \kappa}\right)(1 - s^*),\ A_W = A_Y - A_E,\ B_W = B_Y - B_E. \tag{8.9}$$

The mean values, variances, covariances and correlations of our variables over the detrended cycle are defined by the formulae

$$E(s - s^*) = \frac{1}{T} I_s = m_s, \quad E(x - b^*) = \frac{1}{T} I_x = m_x,$$

$$E(Q_X^*) \overset{Lin}{=} L_X(m_s, m_x) = m_X,$$

$$E[Q_X^* - m_X)]^2 \overset{Lin}{=} \frac{1}{T} \int_0^T L_X^2\, dt - m_X^2 = \sigma_X^2,$$

$$E[Q_X^* - m_X)]\, [Q_Y^* - m_Y)] \overset{Lin}{=} \frac{1}{T} \int_0^T L_X L_Y\, dt - m_X m_Y = \operatorname{cov}(Q_X^*, Q_Y^*),$$

$$r_{XY} = \operatorname{cov}(Q_X^*, Q_Y^*)/\sigma_X \sigma_Y.$$

The relevant functions in comparisons with data and with other models are the σ_X/σ_Y and the correlations r_{XY}. The following formulae are suitable for calculations:

$$\left(\frac{\sigma_X}{\sigma_Y}\right)^2 \overset{Lin}{=} \frac{(\omega/2\pi)\left[A_X^2 I_{ss} + B_X^2 I_{xx} + 2A_X B_X I_{sx}\right] - m_X^2}{(\omega/2\pi) B_Y^2 I_{xx} - m_Y^2}, \tag{8.10}$$

$$r_{XY} \overset{Lin}{=} \frac{(\omega/2\pi)\left[A_X B_Y I_{sx} + B_X B_Y I_{xx}\right] - m_X m_Y}{\sigma_X \sigma_Y}. \tag{8.11}$$

The shortcut formulae obtained for $m_X = m_Y = 0$, i.e.

$$(8.12)\quad \left(\frac{\sigma_X}{\sigma_Y}\right)^2 \approx \left(\frac{A_X}{B_Y}\right)^2 \frac{F_{ss}}{F_{xx}} + \left(\frac{2A_X B_X}{B_Y^2}\right) \frac{F_{sx}}{F_{xx}} + \left(\frac{B_X}{B_Y}\right)^2,$$

$$(8.13)\quad r_{XY} \approx \left[\left(\frac{A_X}{B_Y}\right) \frac{F_{sx}}{F_{xx}} + \frac{B_X}{B_Y}\right] (\sigma_Y/\sigma_X),$$

with the coefficients A and B from (7)-(9) and the F-functions from (4)-(6), do fairly well. A busy reader can use (12)-(13) to check roughly the numerical calculations. Let it be mentioned that the unknown initial state $s(0) - s^*$ cancels out in the formulae (10)-(13).

The prediction formulae, both the (10)-(11) and (12)-(13), include ten parameters. Only six of them are independent of each other. They are chosen to be the parameters $\beta, \kappa, \rho, n, b^*$ and s^*, mainly because they can be given numerical values obtained from classic empirical works or from earlier theoretical usage. Thus the traditional estimates

$$\beta = .25, \quad \kappa = .417, \quad \rho = 2\,\%, \quad n = 1.5\,\%, \quad b^* = .3$$

were chosen: the values of β, n and b^* are either identical or close to the pioneering Denison (1961) estimates based on the averages in the U.S. economy in the period 1909-57, while the number representing κ is the one used by Lucas (1988), who also accepted the value $\rho = .02$ as a realistic one. For the balanced-growth net savings rate s^* the Denison value .1 is too small for the present comparisons, since these comparisons concern a period after the second world war, when the level of the savings rate had risen[1]. Thus three numerical values of s^* will be here experimented with, viz. .13, .15 and .17.

The remaining four parameters are determined by the above six, α and ω as indicated in (3), where the parameter a was given as a function of β and σ. Let it be recalled that the dependence of the risk aversion coefficient σ on the independent parameters is as follows:

$$(8.14)\quad \sigma = \frac{\beta b^* - \rho}{s^* b^* - n}.$$

[1]See N.Mankiw, D.Romer and D.Weil, A contribution to the empirics of economic growth, NBER working paper 3541,1990, but note that their numbers are those of the gross savings rate

This dependence is a consequence of the balanced-growth Euler equation $\rho+\sigma(\lambda-n) = \beta b^*$ and the balanced-growth equation of the growth of physical capital, $\lambda = s^*b^*$. This important parametric relation is valid also in the Lucas (1988) model as well in the Solow model.

Table 7 gives the numerical predictions of correlations with output and the standard deviations, of consumption (C), investment (I), employment (E) and the productivity (Y/E=W). These predictions have been calculated from (10)-(11). To facilitate the reader's checking of the calculations, the numerical values of the dependent parameters, of the mean values and the A- and B-coefficients of the mentioned variables, as well as of the F- and I-functions, such as they have been obtained

TABLE 7. – THE NUMERICAL CALCULATION OF CORRELATIONS AND STANDARD DEVIATIONS

$s^* = .13$	$\alpha = .036$	$\omega = .1035$	$\sigma = 2.2917$
$a = 7.0968$	$F_{ss} = 15.5204$	$F_{sx} = 4.0786$	$F_{xx} = 10.1061$
$m_Y = .0666$	$I_{ss} = 122.5726$	$I_{sx} = 32.2108$	$I_{xx} = 80.5236$
$m_C = ..0291$	$m_I = .3178$	$m_E = .0428$	$m_W = .0238$
$A_C = 0$	$A_I = 0$	$A_E = -.0643$	$A_W = .0643$
$B_C = .1091$	$B_I = 1.1930$	$B_E = .1864$	$B_W = .0636$
$r_{CY} = 1.00$	$r_{IY} = 1.00$	$r_{EY} = .9060$	$r_{WY} = .7654$
$\sigma_C/\sigma_Y = .4364$	$\sigma_I/\sigma_Y = 4.7720$	$\sigma_E/\sigma_Y = .7094$	$\sigma_W/\sigma_Y = .4668$
$s^* = .15$	$\alpha = .03$	$\omega = .0920$	$\sigma = 1.8333$
$a = 8.8$	$F_{ss} = 18.3922$	$F_{sx} = 5.1765$	$F_{xx} = 15.5290$
$m_Y = .0677$	$I_{ss} = 124.2441$	$I_{sx} = 34.9685$	$I_{xx} = 104.9027$
$m_C = .0369$	$m_I = .2419$	$m_E = .0435$	$m_W = .0242$
$A_C = 0$	$A_I = 0$	$A_E = -.0643$	$A_W = .0643$
$B_C = .1364$	$B_I = .6439$	$B_E = .1821$	$B_W = .0679$
$r_{CY} = 1.00$	$r_{IY} = 1.00$	$r_{EY} = .9205$	$r_{WY} = .7951$
$\sigma_C/\sigma_Y = .5455$	$\sigma_I/\sigma_Y = 3.5758$	$\sigma_E/\sigma_Y = .6982$	$\sigma_W/\sigma_Y = .4496$
$s^* = .17$	$\alpha = .024$	$\omega = .0794$	$\sigma = 1.5278$
$a = 11.5789$	$F_{ss} = 22.6934$	$F_{sx} = 6.9753$	$F_{xx} = 26.1567$
$m_Y = .0698$	$I_{ss} = 128.8443$	$I_{sx} = 39.6028$	$I_{xx} = 148.5073$
$m_C = .0457$	$m_I = .1876$	$m_E = .0449$	$m_W = .0249$
$A_C = 0$	$A_I = 0$	$A_E = -.0643$	$A_W = .0643$
$B_C = .1636$	$B_I = .4217$	$B_E = .1778$	$B_W = .0722$
$r_{CY} = 1.00$	$r_{IY} = 1.00$	$r_{EY} = .9395$	$r_{WY} = .8124$
$\sigma_C/\sigma_Y = .6545$	$\sigma_I/\sigma_Y = 2.6867$	$\sigma_E/\sigma_Y = .6841$	$\sigma_W/\sigma_Y = .4208$

and used in calculations, are mentioned for the purpose of comparison. The initial state factor $[s(0) - s^*]$ has been omitted in the mean values m, as well as its square in the I-integrals.

Note on calibration in Chapters 8 and 9:

In Chapter 9 the anomalous business cycles, whose statistics comes from the period 1914-50, will be discussed in terms of the minimal dynamics constructed in Chapter 3. In calculations the anomalous cycles can be approached by starting with the results obtained for the ordinary business cycles. Then the problem is that the statistical records on the ordinary cycles usually concern the second half of the century.

The choice of calibration has been done keeping in mind both of the applications. Therefore the *traditional estimates* of parameters were chosen above. The traditional calibration was used in Section 7 in the calculations concerning the first half of the century. This calibration is not optimal for cycles in the second half of the century. However it proves to be good enough to make the predictions of the minimal dynamics already much better than those of the stochastic models.

The calibration we have called traditional applies some estimates of essential parameters coming from the Denison study (1961) concerning the period 1909-57. These estimates, such as the value $\beta = 1/4$ figured in the main roles still in the standard Samuelson-Nordhaus book ('Economics') in 1985. Later the paper by Mankiw, D.Romer and Weil (1990) recommended the estimate $\beta = 1/3$ for the second half of the century. Still later appeared the Penn Table of Summers and Heston (1991), which for the first time put the problem of aggregate estimates on a firm ground. The Penn Table was mainly used in Chapters 6 and 7, which deal with the decisive evidence in favour of the minimal dynamics.

A problem apart was the choice of the estimate of κ, which for the first time appeared in the Lucas 1988 paper (his γ). The calculations in Chapters 8 and 9 were started soon after the appearance of the Lucas paper, and it was natural to use the estimate $\kappa = .417$ he suggested. From the Penn Table calibrations performed in Chapters 6 and 7 we can see that the correct numbers are much smaller than that.

8.2 Comparisons with Facts and the Stochastic Models of Kydland-Prescott, Hansen-Rogerson and Danthine-Donaldson

Table 8 compares the predictions of Table 7 with the U.S. data and with the predictions of three models based on stochastic shocks, viz. that of Danthine and Donaldson (1990-1993), Hansen and Rogerson (1985), and Kydland and Prescott (their later, 1986 version). The success of each model is expressed by the sum of error squares in each case. The success of the Table 7 models is evident.

TABLE 8. – COMPARISONS OF THE PREDICTIONS OF FIVE MODELS WITH U.S.DATA[2]

Standard deviations:

X :	*The U.S. economy* :	*Table 7* $s^* = .13$:	*Table 7* $s^* = .15$:	*Danthine–Donaldson* :	*Hansen–Rogerson* :	*Kydland–Prescott* :
Y	1.00	1.00	1.00	1.00	1.00	1.00
C	.73	.44	.55	.19	.29	.25
I	4.89	4.77	3.58	3.45	3.24	3.07
E	.94	.71	.70	.72	.77	.68
W	.67	.47	.45	.35	.28	.40
$\sum\Delta^2$:		.1961	1.8645	2.5185	3.1071	3.6933

Correlations with output:

X :	*The U.S. economy* :	*Table 7* $s^* = .13$:	*Table 7* $s^* = .15$:	*Kydland–Prescott* :	*Hansen–Rogerson* :	*Danthine–Donaldson* :
C	.85	1.00	1.00	.85	.87	.69
I	.92	1.00	1.00	.88	.99	.99
E	.76	.91	.92	.95	.98	.98
W	.42	.77	.80	.86	.87	.91
$\sum\Delta^2$:		.1739	.1989	.2313	.2562	.3190

[2]The numbers other than those in the BBC columns are taken or computed from those given by Danthine and Donaldson (1993) who, for Hansen-Rogerson and Kydland-Prescott models, quoted Prescott (1986). The last row indicates the sum of error squares for each model.

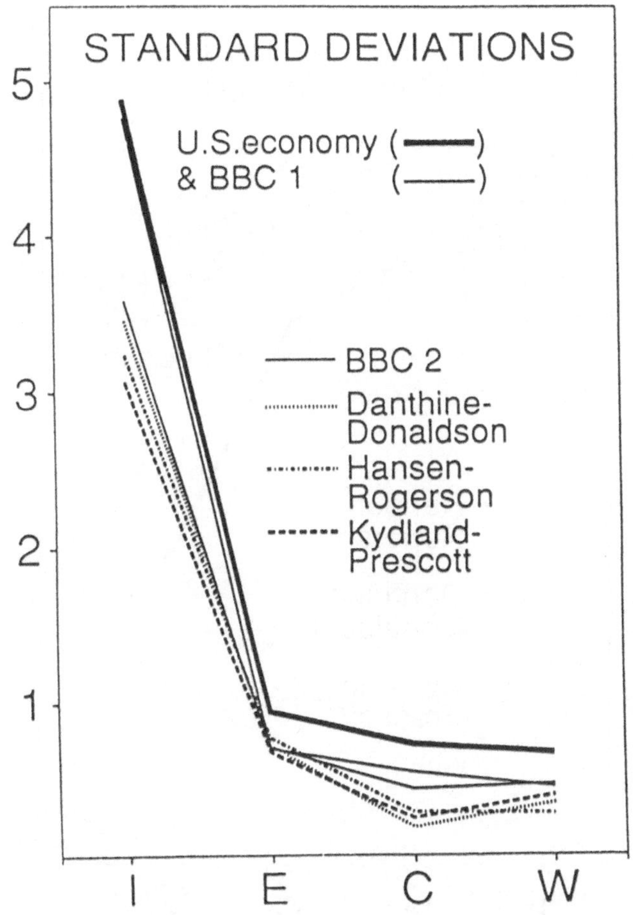

FIGURE 9.

Figures 9 and 10, which illustrate the numbers shown in Table 8, confirm the impression given by the numbers and indeed add something to it: what makes the difference is that the predictions of this theory follow the *pattern* of empirical correlations and standard deviations better than do the predictions derived from the stochastic models.

Table 9 on page 133 shows a similar comparison of two Table 7 versions, with $s^* = .15$ and with $s^* = .17$, with the original Kydland-Prescott model and with the data they used (Kydland and Prescott,1982).

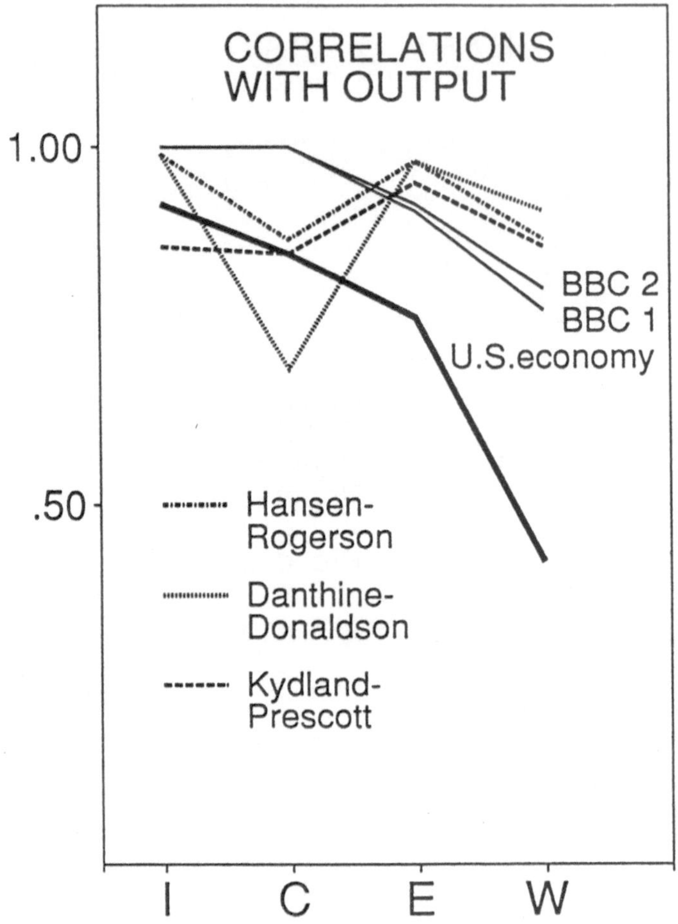

FIGURE 10.

Again the non-stochastic Table 7 models are better in their predictions.

Taken together Tables 8 and 9 show that the Table 7 version with $s^* = .15$ is the best when both comparisons are considered. The version with $s^* = .13$ takes best into account the peculiarities in the U.S. economy as displayed by the material in Table 8, while the version with $s^* = .17$ is the most successful in the comparisons shown in Table 9 .

Scarce though the material is in these comparisons, I have deliber-

TABLE 9. – COMPARISONS WITH THE KYDLAND-PRESCOTT (1982) MODEL AND WITH THE U.S. ECONOMY IN THE PERIOD 1950-79[3]

Standard deviations:

X:	*The U.S. economy*:	*Table7* 3 $s^* = .17$:	*Table7* 2 $s^* = .15$:	*Kydland–Prescott*:
Y	1.00	1.00	1.00	1.00
C	.33	.65	.55	.50
I	2.83	2.69	3.58	3.58
E	1.11	.68	.70	.58
W	.55	.42	.45	.50
$\sum\Delta^2$:		.3238	.7890	.8748

Correlations with output:

X:	*The U.S. economy*:	*Table7* 2 $s^* = .15$:	*Table7* 3 $s^* = .17$:	*Kydland–Prescott*:
C	.94	1.00	1.00	.66
I	.71	1.00	1.00	.80
E	.85	.92	.94	.93
W	.10	80	.84	.90
$\sum\Delta^2$:		.5826	.6434	.7329

ately made it even scarcer by leaving out capital (Lucas,1987, also discarded it in his Kydland-Prescott Table). The problem with quantitative comparisons involving the dynamics of capital still comes from the fact that the capital stock, which can be measured, is not the same thing as the capital input to production, i.e. the capital services, which is relevant in dynamics. One can easily understand that capital stock does not vary much during a cycle, and that it has almost a zero correlation with output over a cycle. But 'capital' in growth theory is a different thing, whose reliable measurement is next to impossible. It is therefore excluded from the comparisons in this book.

[3]The data and the predictions of the Kydland-Prescott model as given in Kydland and Prescott (1982). The last row indicates the sum of error squares in each model.

8.3 Comparisons with Empirical Autocorrelations

The autocorrelation of output Y with its value Y' delayed by a time D is given by

$$r_{YY'} = \mathrm{cov}\,(Y,Y')/\sigma_Y^2, \quad \mathrm{cov}\,(Y,Y') = I_{YY'} - m_Y m'_Y$$
$$I_{YY'} = \frac{1}{T}\int_0^T Q_Y^*(t) Q_Y^*(t-D)\,dt,$$

$$m'_Y = \frac{1}{T}\int_0^T Q_Y^*(t-D)\,dt, \quad \sigma_Y^2 = \mathrm{cov}\,(Y,Y') \text{ for } D = 0.$$

Here $m'_Y \neq m_Y$ because of the divergence of the cycles defined by (1) and (2), which we shall use. Direct calculation gives, by applying (3.85) and (2):

$$(8.15)\, I_{Y,Y'} = \frac{1}{32}\left[\frac{\beta a\left(\alpha^2+4\omega^2\right)}{(1-s^*)\omega}\right]^2 \left(e^{2\pi\alpha/\omega}-1\right) e^{-\alpha D/2}\,.$$
$$\left[\frac{\cos\omega D}{\alpha} - \frac{\alpha\cos\omega D - 2\omega\sin\omega D}{\alpha^2+4\omega^2}\right][s(0)-s^*]^2,$$

$$(8.16)\quad m'_Y/m_Y = \frac{e^{(2\pi/\omega - D)\alpha/2}\left(\cos\omega D + \frac{\alpha}{2\omega}\sin\omega D\right) - 1}{e^{\pi\alpha/\omega}-1}.$$

Again the unknown initials state $s(0) - s^*$ cancels out in the final formulae.

In a comparison of the predictions of the dynamics constructed in Chapter 3 with those of the Kydland-Prescott (1982) model, which is the classic case of the comparison of a shock model with autocorrelation data, the theoretical equivalent of a quarter of a year is needed. By taking the average length of period of the business cycles in the U.S. economy in the period 1950-79, studied by Kydland and Prescott, to be 4 years, we have $g_o = T/4$, where $T = 2\pi/\omega$ is the length of period in terms of the theoretical unit of time [TU] (for the time scales, see Section 3.2). Thus $D_1 = T/16 = \pi/8\omega$ is the theoretical equivalent of a quarter of a year. Numerical calculations are given in Table 10.

TABLE 10. – THE CALCULATION OF AUTOCORRELATIONS[4]

$s^* = .11$	$j=1$	$j=2$	$j=3$	$j=4$	$j=5$	$j=6$
$D = jD_1$	3.4563	6.9126	10.3690	13.8253	17.2816	20.7379
$e^{-\alpha D/2}$	.9300	.8649	.8043	.7480	.6959	.6469
$\cos \omega D$	.9239	.7071	.3826	0	−.3826	−.7071
$\sin \omega D$	.3826	.7071	.9239	1	.9239	.7071
$m_Y m'_Y$	.0040	.0027	.0009	−.0011	−.0030	−.0045
$\text{cov}\,(YY')$	.0652	.0515	.0324	.0115	−.0080	−.0234
$r_{YY'}$	.9272	.7326	.4612	.1632	−.1142	−.3331
$s^* = .15$	$j=1$	$j=2$	$j=3$	$j=4$	$j=5$	$j=6$
$D = jD_1$	4.2674	8.5348	12.8023	17.0697	21.3371	25.6045
$e^{-\alpha D/2}$	.9380	.8798	.8253	.7741	.7261	.6811
$\cos \omega D$	.9239	.7071	.3826	0	−.3826	−.7071
$\sin \omega D$	.3826	.7071	.9239	1	.9239	.7071
$m_Y m'_Y$	.0040	.0026	.0006	−.0017	−.0038	−.0054
$\text{cov}\,(YY')$	.0848	.0669	.0417	.0138	−.0124	−.0333
$r_{YY'}$	.8957	.7065	.4403	.1456	−.1311	−.3514
$s^* = .18$	$j=1$	$j=2$	$j=3$	$j=4$	$j=5$	$j=6$
$D = jD_1$	5.4162	10.8323	16.2485	21.6647	27.0808	32.4970
$e^{-\alpha D/2}$	.9447	.8925	.8432	.7965	.7525	.7110
$\cos \omega D$	.9239	.7071	.3826	0	−.3826	−.7071
$\sin \omega D$	.3826	.7071	.9239	1	.9239	.7071
$m_Y m'_Y$	.0054	.0035	.0007	−.0024	−.0053	−.0076
$\text{cov}\,(YY')$	.1185	.0933	.0576	.0178	−.0198	−.0500
$r_{YY'}$	.9197	.7241	.4471	.1386	−.1537	−.3879

The comparison with the Kydland-Prescott (1982) model and with the data from the U.S. economy in the period 1950-79, which they used, are given in Table 11. Three versions of the dynamics constructed in Chapter 3 were again experimented with, this time determined by the values .11, .15 and .18 of the balanced-growth net savings rate s^*. As shown by the sums of error squares reported in Table 11 the value $s^* = .15$ gives again the best prediction of data. But also the other versions of this dynamics predict the data better than the stochastic model. Again also the pattern is better.

[4]The initial state factor $[s(0) - s^*]^2$ has been deleted in $m_Y m'_Y$ and in $\text{cov}\,(YY')$. It is canceled out in $r_{YY'}$. Note that $\omega D = (j/16)360°$.

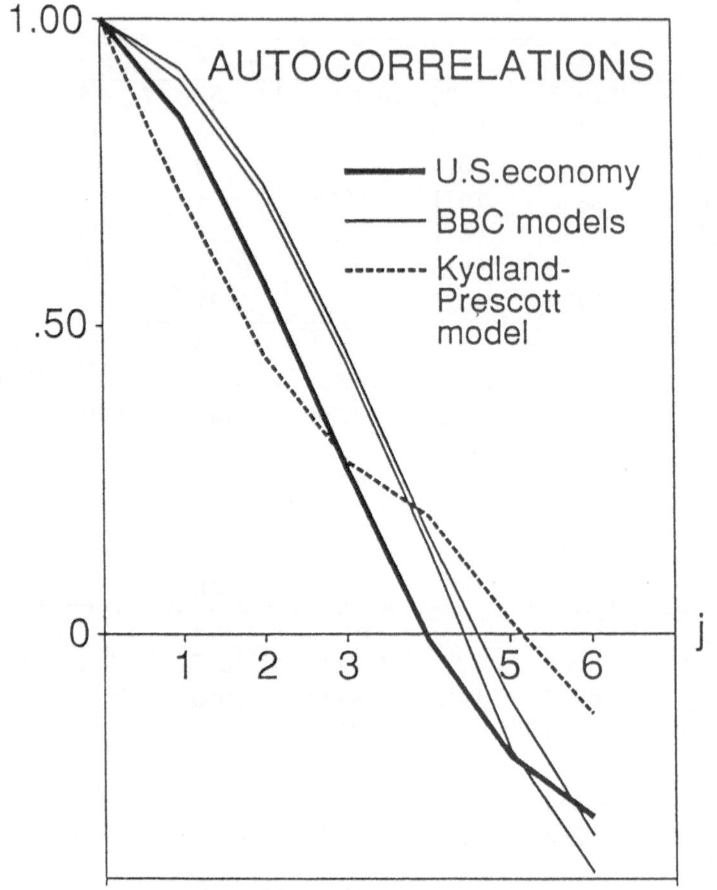

FIGURE 11. – THE AUTOCORRELATIONS OF OUTPUT COMPUTED IN TABLE 10 COMPARED WITH THE 1982 DATA OF KYDLAND-PRESCOTT AND WITH THEIR ORIGINAL PREDICTIONS.

Later Kydland and Prescott[5] corrected their data on autocorrelations of output and published the following numbers:

$$.85\ (j=1),\ .63\ (j=2),\ .38\,(j=3),\ .16\ (j=4),\ -.02\ (j=5).$$

[5]Kydland,F. and E.C.Prescott, Business cycles: real facts and monetary myth, Federal Reserve Bank of Minneapolis Quarterly Review 14, 1990, p.3-18.

TABLE 11. – COMPARISONS OF PREDICTED AUTOCORRELATIONS WITH THOSE OF THE KYDLAND-PRESCOTT MODEL AND WITH DATA[6]

j :	*The U.S. economy* :	*Table* 10 $s^* = .15$:	*Table* 10 $s^* = .18$:	*Table* 10 $s^* = .11$:	*Kydland–Prescott* :
1	.84	.90	.92	.93	.71
2	.57	.71	.72	.73	.45
3	.27	.44	.45	.46	.28
4	−.01	.15	.14	.16	.19
5	−.20	−.13	−.19	−.11	.02
6	−.30	−.35	−.39	−.33	−.13
$\Sigma\Delta^2$:		.0851	.0951	.1077	.1487

With these numbers both the Kydland-Prescott predictions and those of the macroeconomic dynamics of Chapter 3 do somewhat better, while the order of their success remains the same. The sum of error squares of the former predictions is now .0645 while the minimal dynamics of this book now has a number as .0447. Or, to take the averages of these numbers, since now there are five instead of six delays:

– Kydland-Prescott model has the average prediction success .0129 instead of the earlier value .0248.

– The minimal dynamics has the average prediction success .0089 instead of the earlier .0142.

[6]The data and the Kydland-Prescott predictions are here those reported in Kydland and Prescott, 1982, and again in Lucas, 1987. The last row indicates the sum of error squares for each model.

8.4 Which Predicts Better the Data: The Latest Standard Stochastic Model (Cooley-Prescott) or the Minimal Macroeconomic Dynamics?

After the above calculations were performed some years ago Thomas Cooley and Edward Prescott constructed, in connection with the first 'Frontiers'-book concerning the business cycle[7] a kind of standard model of stochastic growth economy on the basis of all the knowledge so far accumulated of such models based on stochastic optimization.

The building of this particular model was very careful: "The model has been simulated 100 times, with each simulation being 150 periods long, to match the number of observations underlying the statistics reported in Table 1.1 [here Table No. 12 – A.A.]. The simulated data were filtered by using the H-P filter just as the original data were to give us the same representation of the business cycle. Table 1.2 [here also in Table No.12 – A.A.] presents the standard deviations of the key variables from the model economy. In addition we present the cross-correlation of each of the variables with output." (Thomas F. Cooley and Edward C. Prescott, Economic growth and business cycles, Thomas F. Cooley (ed), ibid., p.33.)

The authors go on by asking:

1) "What do we learn from this exercise? One question that is answered is: How much of the variation in output can be accounted for by technology shocks? In this artificial economy, output fluctuates less than in the U.S. economy, suggesting that much, but not all, of the variation of output is accounted for by technology shocks."

2) "The labor input [referring to 'Hours'] in this model economy fluctuates only about half as much as in the U.S. economy suggesting that some important feature of the labor market is not captured here."

3) "Hours and productivity in the model economy go up and down together, whereas in the data they do not. This also suggests an important missing element in this artificial economy as a business cycle model."

---

[7]Thomas F. Cooley (ed.), Frontiers of Business Cycle Research, Princeton University Press 1995.

4) "Investment in the model economy fluctuates much more than does output, just as it does in the U.S. economy."

5) "Consumption in the model economy fluctuates much less than does output and less than consumption of nondurables and services in the U.S. economy."

6) "Consumption, investment, and the hours in the model economy are all strongly procyclical, as they are in the U.S. economy." [The citations 1-6 are from the text on p.33 in the book in question. – A.A.]

Table 12 gives the empirical results from the U.S. Economy in the period 1954-91 as used by Cooley and Prescott (column 1), their Cooley-Prescott theoretical predictions (column 2), and the theoretical predictions obtained from the minimal dynamics (column 3) as given in Table 7 before – again the model $s^* = .13$ is used.

We can see that the minimal macroeconomic dynamics is superior in the prediction of both the standard deviations and correlations.

TABLE 12. – THE PREDICTIONS OF COOLEY-PRESCOTT MODEL COMPARED WITH THE MINIMAL MACROECONOMIC DYNAMICS.

Standard deviations:

Variable	*U.S. economy*	*Cooley – Prescott*	*Table 7*
Output	1.00	1.00	1.00
Consumption	.73	.24	.44
Investment	4.79	4.41	4.77
Hours	.98	.57	.71
Productivity	.42	.45	.47
$\sum \Delta^2$:		.5535	.1599

Correlations with output:

Variable	*U.S. economy*	*Cooley – Prescott*	*Table 7*
Consumption	.83	.84	1.00
Investment	.91	.99	1.00
Hours	.92	.99	.91
Productivity	.34	.98	.77
$\sum \Delta^2$:		.4210	.2220

As to the point 1, the question will be answered in Chapter 10, where it will be shown that when the technological shocks are superposed on the ordinary business cycles of the minimal macroeconomic dynamics constructed in Chapter 3, only 10 percent of the variance of output is explained by them, while 90 percent is explained by the strictly causal minimal dynamics.

The points 4 and 6 are well predicted both by the Cooley-Prescott stochastic model and by the minimal dynamics of this book.

In the points 2,3 and 5 a radical difference exists between prediction success of the Cooley-Prescott model and that of the minimal macroeconomic dynamics, in favour of the latter.

The result is rather clear: the Cooley-Prescott model has the total sum of error squares .9745, while the (linear approximation of) the minimal macroeconomic dynamics has only .3819.

Although the evidence given in this chapter can be considered only complementary to the decisive evidence given in Chapters 6 and 7 above, the goodness of the minimal macroeconomic dynamics also in the prediction of correlations and standard deviations over detrended cycles is remarkable, especially when compared with the rather weak results of the popular stochastic models.

8.5 Business Cycles of Total Working Hours

The use of human time, as analysed in Chapter 3, is a little more complicated than what has been usually considered in economics. This is mainly because the use of leisure has been given new economic relevance. It follows among other things that the variable called 'total working hours' and marked by H comprises all the hours outside the leisure time. Therefore its cycle function and business cycle deserves a special treatment.

The Cycle Function of Total Hours Defined

The 'total working hours' H of the working-age population has been further decomposed in two different ways to a sum, viz. 1) to the sum of the hours in physical work H_{phys} devoted to the production of output and the hours of intellectual work H_{int}, and 2) to the sum of the paid working hours H_{paid} and the hours H_{ed} spent in formal education:

$$(8.17) \qquad H = H_{phys} + H_{int} = H_{paid} + H_{ed}.$$

What is measurable and indicated in economic statistics (in the U.S. at least) is the paid working hours H_{paid}. On the other hand, the hours of physical work in production H_{phys} and the hours in formal education H_{ed} are both of them variables with no business cycles, i.e. their cycle functions are

$$(8.18) \qquad Q_{H_{phys}} = Q_{H_{ed}} = 0.$$

Therefore the cycle function of the 'total hours' H is equal to that of the hours in intellectual work H_{int}, which in turn is equal to the cycle function of the paid working hours H_{paid}:

$$(8.19) \qquad Q_H = Q_{H_{int}} = Q_{H_{paid}}.$$

It is actually the paid working hours that are given in economic statistics, as was mentioned above.

We can best calculate, of the three mutually equal cycle functions (19), the cycle function of the hours in intellectual work. We get from the definition of H_{int} in (3.3) and from the growth equation of human capital (3.10) that

$$H_{int} = \frac{\dot{h}/h}{k}. \tag{8.20}$$

By the way, note that by reading this equation the other way round we can see the meaning of the function k:

$$k = \frac{\dot{h}/h}{H_{int}}.$$

The larger is the growth rate of human capital and the smaller is the time used for attaining this growth rate, the larger is k. Thus k indicates the speed of learning knowledge. The level factor k_o in the function $k = k_o e^{mt}$ indicates the level of this speed, which may change from one country to another one and from a historical period to another one. It can be expected to be higher in more advanced countries. Obviously the length of the time H_{int} devoted to learning knowledge in a country depends on the level of the speed of learning knowledge in that country.

By using the equations (3.83),(5.5) and $k = k_o e^{mt}$ we get (20) expressed in the form

$$H_{int} = \frac{\nu}{k} + \frac{e^{-mt}\, Q_h}{k_o}, \quad \text{thus} \quad Q_{H_{int}} = \frac{e^{-mt} Q_h}{k_o}. \tag{8.21}$$

Since H and H_{int} have the same cycles we have, in view of (3.7), (3.19) and (5.4), the following extraordinary equation separating its value on the balanced-growth path from its cycle function:

$$H = \frac{\alpha^* + \nu}{k} + \frac{e^{-mt}\, Q_h}{k_o} = (H)_P + Q_H. \tag{8.22}$$

Calculation of Output Correlation and SD

Following the method of earlier calculations we shall use for Q_h its linearized form, taking the coefficients $A_h = A_E$ and $B_h = B_E$ from (8) and (9), respectively. This gives:

$$Q_H = \frac{1}{k_o}\left(\frac{\beta}{1-\beta+\kappa}\right) e^{-mt} \left[(1-s^*)(x-b^*) - b^*(s-s^*)\right]. \tag{8.23}$$

For the standard deviation proportion σ_H/σ_Y and the correlation r_{HY} we can apply the following short formulae corresponding to (12) and (13):

$$(8.24)\quad (k_o)^2\left(\frac{\sigma_H}{\sigma_Y}\right)^2=\left(\frac{1}{1-\beta+\kappa}\right)^2\left[(b^*)^2\frac{F'_{ss}}{F'_{xx}}-2b^*(1-s^*)\frac{F'_{sx}}{F'_{xx}}+(1-s^*)^2\right],$$

$$(8.25)\quad r_{HY}=\frac{1}{k_o}\left[-\left(\frac{b^*}{1-\beta+\kappa}\right)\frac{F'_{sx}}{F'_{xx}}+\frac{1-s^*}{1-\beta+\kappa}\right]\frac{\sigma_Y}{\sigma_H}.$$

Note that here F' does not indicate derivation but only a function distinguished from the F–functions in (12)-(13).

The functions F' are as follows:

$$(8.26)\quad F'_{ss} = \frac{(\alpha^*)^2+4\omega^2}{8\omega^2(\alpha-2m)}+ + \alpha^*\left[\frac{(3/2)-(\alpha^*)^2/8\omega^2-\alpha^*m/4\omega^2-m/\alpha^*}{(\alpha^*)^2+4\omega^2+4m^2-4\alpha^*m}\right],$$

$$(8.27)\quad F'_{sx} = \frac{a\left((\alpha^*)^2+4\omega^2\right)}{16(1-s^*)\,\omega^2}\cdot \cdot\left[\frac{4\omega^2-(\alpha^*)^2+2\alpha^*m}{4\omega^2+(\alpha^*)^2+4m^2-4\alpha^*m}+\frac{\alpha^*}{\alpha^*-2m}\right],$$

$$(8.28)\quad F'_{xx} = \frac{a^2\left((\alpha^*)^2+4\omega^2\right)^2}{8(1-s^*)^2(\alpha^*-2m)\left((\alpha^*)^2+4\omega^2+4m^2-4\alpha^*m\right)}.$$

These functions of course reduce to the corresponding F-functions, given in (4)-(6), respectively, for $m=0$.

Predictions Compared with Facts

The calibration follows the traditional calibration applied before in this chapter:

$$s^*=.13,\quad b^*=.30,\quad \alpha^*=.036,\quad \beta=.25, \sigma=2.2917,$$

$$\Longrightarrow\ a=50.363068,\quad \omega=.1035.$$

The value $m=.05$ obeying the condition (18.24) was chosen. With these values for the parameters we get:

$F'_{ss} = -8.0038651, \quad F'_{sx} = .8389708, \quad F'_{xx} = -5.3946384,$
after which we can easily compute the following results:

$$\text{(8.29)} \qquad \frac{\sigma_H}{\sigma_Y} = \frac{.8446473}{k_o} \quad \text{and} \quad r_{HY} = .9299507 \approx .93.$$

The predicted correlation between 'total hours' and output, .93, is in agreement with Finn Kydland's general statement[8], according to which the empirical contemporaneous correlation coefficients of 'total hours' with real GNP are near .90. Indeed the empirical values given by him from the period 1954-91 in the U.S. economy are .88 (Establishment Survey) and .86 (Household Survey), while Cooley and Prescott give for the same period in the same country the correlations .92 (Establishment Survey) and .86 (Household Survey). Obviously the theoretical prediction is of the correct magnitude, and we have here again a prediction success of the macroeconomic dynamics constructed in Chapter 3.

As to the predicted standard deviation proportion σ_H/σ_Y, it depends on the level k_o of the speed of learning knowledge, and we have here a method of estimating this level in the U.S. as an average of the period 1954-91. If the Establishment Survey is used, the empirical values of this proportion, viz. .98 (Cooley-Prescott) and .96 (Kydland), give the mutually compatible results

$$k_o = .862 \quad \text{and} \quad k_o = .880,$$

respectively. The Household survey results ($\sigma_H/\sigma_Y = .87$) on the other hand suggests a little higher value

$$k_o = .971$$

in both cases.

Note (Improved predictions in Table 8.) The empirical values used in Table 8 for the employment variable E were in fact measured for 'total hours', while the predictions given by the minimal dynamics were

[8]In Thomas F. Cooley (ed.), ibid., p. 125.

computed for the employment E. If we replace the 'total hours' empirics by the empirical values

(8.30) $$\left(\frac{\sigma_E}{\sigma_Y}\right)^{emp}_{1954-91} = .63 \text{ (Kydland), or } .82 \text{ (Cooley-Prescott)}$$

(8.31) $$(r_{EY})^{emp}_{1954-91} = .83 \text{ (Kydland) or } .89 \text{ (Cooley-Prescott),}$$

given by Kydland, on the one hand, and Cooley and Prescott, on the other, the predictions of the minimal dynamics, calculated in Table 7 and removed to Table 8, are clearly improved.

Accepting the Kydland reading of the period 1954-91 in the U.S. economy the errors in the predictions of the minimal dynamics are only $\Delta = .08$ in both cases, instead of the errors $\Delta = .23$ (for the standard deviation proportion) and $\Delta = .15$ (for the correlation) reported in Table 8. Therefore the square sums of errors of these predictions become

(8.32) $$\sum \Delta^2 = .1496 \quad \text{instead of } .1961,$$

for standard deviations, and

(8.33) $$\sum \Delta^2 = .1578 \quad \text{instead of } .1739$$

for correlations.

The predictions of the three stochastic models were also computed for 'total hours', and accordingly have no corresponding corrections.

8.6 Comments and Challenges

Relative Independence of Predictions from Calibration

One of the critical remarks one can make concerning the calculations performed in this chapter concerns the applied calibration. The 'traditional' calibration, chosen here for the reasons explained in Section 8.1, is not optimal for the calculations of Chapter 8.

Still the theoretical predictions obtained also in this chapter have been superior to those given by the conventional stochastic models, when compared with the basic empirical facts. Indeed there seems to be a relative independence from the chosen calibration in the predictions of the theory constructed in Chapter 3. We have seen this in special in Chapters 6 and 7, in which an invariant pattern appeared in the results from one calibration to another. The theory predicted correctly Plosser effect (in all the four cases studied) and its quantitative extensions (in three cases out of four) with widely different calibrations obtained from economic statistics in different countries. The theory also predicted correctly the invariant pattern appearing in the leads and lags in three different periods. The result also confirmed the expectation of the theory that the two periods from the second half of the century were very close to one another, while the leads and lags in the first half of the century were characteristically shorter.

The good results in the prediction of empirical facts in this chapter seem to confirm the relative independence of theoretical predictions from calibration.

Two Levels of Macroeconomic Theory

The linear approximation applied Part 2 certainly does not full justice to the theory of Part 1, neither to the nonlinear cycle equations nor to the many nonlinear cycle functions. Still its predictions are better than those of the stochastic shock models in all the quantitative comparisons performed. This non-stochastic theory also gives correctly all the qualitative patterns of data in Figures 3 and 6-11, which cannot be said of any of the considered principal stochastic models including the latest standard model of Cooley and Prescott.

The conclusion is near that the dynamics based on the canonical formalism of macroeconomics constructed in Chapter 3 reflects fundamental characteristics of the real dynamics of economic growth and the business cycle, fundamental enough to be taken into account in the theory of long-term economic development.

On the other hand, we know that stochastic shocks do exist and affect human life in most of its aspects, and of course they influence economic development and the business cycles. But the success of the canonical formalism suggests that the effects stochastic shocks on business cycles should better be seen as disturbances superposed on the causal cycles than as the factors dominating the business cycle. An example of such a superposition will be given in Chapter 10, but this example is not the only possibility of superposition of stochastic shocks on causal cycles.

Then macroeconomic theory should have to include two levels:

– a short-term level where material values dominate and the theory is reducible to microeconomics, for instance in terms of the Bellman-Lucas formalism in Lucas (1987) with its finite stochastic value functions, also emphasized in Cooley (1995), and

– a long-term level, where the human pursuit of nonmaterial values of ever greater individual freedom and increasing objective knowledge set the tone, and the economic theory accordingly cannot be reducible to microeconomics. The minimal macroeconomic dynamics constructed in Chapter 3 belongs to this level and shows the interaction between material and nonmaterial values as the ultimate source of economic growth including the business cycles.

Let it be mentioned that at least according to the linear approximation of this theory, as applied in Part 2, the relative significance of the business cycle with respect to output diminishes with time, approaching asymptotically zero. It follows from (3.46)-(3.47) as well as from (3.52)-(3.53) that

$$(8.34)\quad \sigma_Y/Y^* \sim e^{(\alpha/2-\lambda)t} \longrightarrow 0 \text{ with } t \to 0, \text{ for } s^* > \beta/3.$$

The condition holds good in all advanced economies.

Thus the business cycles and with them the difference between the two levels, suggested by this theory, may vanish with time.

Convergent instead of Divergent Cycle Functions?

The prospect of the vanishing proportional significance of the business cycle evokes the further question whether the detrended business cycle would be more accurately represented by the converging functions $e^{-\lambda t}[s(t)-s^*]$ and $e^{-\lambda t}[x(t)-b^*]$,from $t=0$ to $t=T$, than by the diverging functions $s(t)-s^*$ and $x(t)-b^*$.

With the convergent functions we would get, for the parameter values $s^* = .13$ and $b^* = .3$, the following results, to be compared with the US values and the predictions of this theory given in Table 8):

$r_{CY} = 1.00$	$r_{IY} = 1.00$	$r_{EY} = .93$	$r_{WY} = .33$
$\sigma_C/\sigma_Y = .44$	$\sigma_I/\sigma Y = 4.77$	$\sigma_E/\sigma_Y = .81$	$\sigma_W/\sigma_Y = .74$

The sum of error squares for the prediction of correlations would be only .0659, which is much smaller than the corresponding total error .1739 obtained for the divergent cycles in Table 8. In the case of the standard deviation proportions the sum of error squares would be .2188. The total sum of error squares would thus for the convergent (but still linearized) cycle functions be .2847 instead of the number .3700 obtained with the divergent cycles in Table 8.

9 Correlations over Anomalous Business Cycles

(*Complementary Evidence*)

9.1 Why Anomalies Are Important in Science?

In 1992 a group of economists published an interesting article[1], in which they observed anomalous business cycles in the period 1914-50 in the U.S.A. and the U.K. Their results can be summarized as follows. Characteristic of this period in these countries was

– that the normally highly procyclical consumption and investment lost their procyclicality;

– that the fall in the procyclicality of investment was still larger than that in the procyclicality of consumption, the correlation with output over detrended business cycles being reduced in the case of investment to .16 (U.S.) or to –.41 (U.K.), and in the case of consumption to .51 (U.S.) or to –.33 (U.K.);

– that as a contrast to these anomalous phenomena employment retained its usual high procyclicality, its correlation with output over detrended cycles being .78 in the U.S. economy and .92 in the U.K. economy; and

– that in the U.K. economy productivity, usually moderately procyclical or neutral, turned into a highly anticyclical variable, having a correlation as low as –.61 with output over detrended cycles. (Productivity did not belong to their U.S. data.)

There is an aspect of the theory of Chapter 3 that makes it more suitable for discussions of anomalous phenomena than are the conventional economic models. In the construction of an economic model usually a great number of specific hypotheses pertaining to different fields of economics are brought together. The result is a sophisticated and more or less balanced combination of various aspects of current economic thinking: often a sort of compendium of the economic wisdom of the day. But this implies, methodologically, that only what is

[1] Isabel H. Correia, Joao L. Neves and Sergio Rebelo, Business cycles from 1850 to 1950, European Economic Review 36, 1992, p.459-467.

average, ordinary and simplest to the common reason can be covered by the models thus constructed.

What is exceptional and anomalous is easily ignored in this kind of model approach, which involves numerous hypotheses, each of which must sound reasonable enough to satisfy common sense. Still the anomalous too may be methodical – it only follows a logic different from the obvious one. Such anomalous cases indeed play a remarkable role for instance in the selection of correct physical theories from the wrong ones. They have often offered the essential arguments for or against a general theory. The history of science knows many examples of this. The small deviations from the perfect circular symmetry in the orbits of planets settled the dispute in favour of the Keplerian theory and against the Copernican one. The small anomalies in the perihelic motion of Mercurius testified for the general theory of relativity, etc.

The theory of macroeconomic dynamics constructed in Chapter 3 was of the type of a general theory, not a 'model'. It was based on a generalization of the macroeconomic value function by introducing just one new term in that function, but avoiding any other hypotheses, including the reducibility to microeconomics and the conventional stochastic form called in Section 1.2 the 'dogmas' of conventional macroeconomics. The result was called a *minimal* macroeconomic dynamics just because the basic macroeconomic facts 1-17 were to be explained by means of the least possible number of hypotheses. The methodical idea was to derive as much as possible from as little as possible. This is somewhat quite contrary to what is involved in the typical construction of an economic model: there many hypotheses are used to derive often not so many consequences readily comparable to empirical facts. The current stochastic models of the business cycle are a case in point.

It was due to the generalized value function that an anomalous growth type appeared in the theory constructed in Chapter 3. The Solow model and the Lucas growth theory are both of them examples of theories based on a balanced-growth path, and thus meant to explain what is obvious and average. Only the generalization of the macroeconomic value function created also an anomalous growth type that obeys a logic similar to that obeyed by the observed anomalous business cycles. The difference from the ordinary cycles is that now the cycle center P is not a fixed point on the (s,x)–plane but moves slowly upwards in the positive direction of the variable x. The cycles are changed accordingly, and this produces, as we shall see, something very much like the observed anomalous correlations over detrended cycles.

We shall first construct a method of calculation useful in the applications of the minimal macroeconomic dynamics to anomalous business cycles. After that a specific model for the present purpose will be chosen within the general framework of the theory. Then the basic macroeconomic fact 17, involving the anomalies mentioned above in output correlations, will be discussed in terms of this theory.

Approximation of Anomalous Cycle Functions

We can again start with the solutions of the linearized equations of the ordinary cycles, valid in a neighbourhood of the fixed cycle center P and expressible in the form (8.1)-(8.2). Again the direction of the flow, whether clockwise or counterclockwise, is irrelevant because of the symmetry of the corresponding solutions. We shall reduce the anomalous cycle functions, by using in a first approximation a loglinear trend, to the ordinary cycle functions of Section 3.4 added by a change variable.

Let V_X denote the resulting approximate anomalous cycle function. By writing $b = b^* - (b^* - b)$ and using the symbol $b^* - b = \Delta$ we can reduce each anomalous cycle function V_X, representing the anomalous cycles detrended over a loglinear trend, to a sum

$$(9.1) \quad V_X = Q^*_X + \Delta_X. \qquad \text{(Rule 1)}$$

Here

$Q^*_X =$ the linearized ordinary cycle function of the economic variable X, with the coefficients (8.7)-(8.9), and

$\Delta_X =$ a *change variable* indicating how the now moving cycle center P affects the cycle.

Thus change variable is obviously given by

$$(9.2) \quad \Delta_X = \left[\left(\frac{\dot{X}}{X}\right)_{P_A} - \left(\frac{\dot{X}}{X}\right)_{P_O} \right] + Q_X(s - s^*, \Delta), \quad \text{(Rule 2)}$$

where $Q_X(s - s^*, \Delta)$ is obtained from the corresponding cycle function $Q_X(s, x)$ of Section 3.4 by writing it first as a function of the variables

$s - s^*$ and $x - b$, and by replacing the latter by Δ. Here P_A and P_O are the cycle centers of the anomalous and ordinary cycles, respectively.

For V_Y and V_C we get at once:

$$\begin{aligned} (9.3) \qquad V_Y &= Q_Y^* + \Delta_Y, \text{ where} \\ & Q_Y^* = \beta(x - b^*) \text{ and } \Delta_Y = (\beta - \beta/\sigma)\Delta, \\ (9.4) \qquad & V_C = Q_C^* = (1/\sigma)Q_Y^*, \ \Delta_C = 0. \end{aligned}$$

To get V_I we first expand $(1 - s)/s$ in a series by the Cauchy rule:

$$\begin{aligned} \frac{1-s}{s} &= \frac{1 - s^* - (s - s^*)}{s^*\left[1 + \frac{s - s^*}{s^*}\right]} = \left[\frac{1 - s^*}{s^*} - \frac{s - s^*}{s^*}\right]\left[1 - \frac{s - s^*}{s^*} + \ldots\right] \\ (9.5) \qquad &= \frac{1 - s^*}{s^*} - \frac{s - s^*}{(s^*)^2} + \text{higher terms.} \end{aligned}$$

The infinite series in $s - s^*$ converges strongly for very small values of $s(0) - s^*$. As we shall be interested only in such initial states very near to the point P_O, the above two first terms will be sufficient for the present purposes. The resulting anomalous cycle function V_I of investment is now easily constructed according to the Rules 1 and 2. It can be expressed, after some calculation, in the following useful form:

$$\begin{aligned} (9.6) \ V_I &= (F - G)Q_Y^* - F^2(\beta - \beta/\sigma)(s - s^*)(x - b^*) + F\Delta_Y \\ (9.7) \qquad &- F^2(s - s^*)\Delta_Y, \ F = \left[\frac{1}{(s^*)^2}\right], \ G = \frac{1}{\sigma}\left(\frac{1 - s^*}{s^*}\right). \end{aligned}$$

Finding the anomalous cycle function V_E of employment offers no problems. First we write $x = (x - b^*) + b^*$ and $1 - s = 1 - s^* - (s - s^*)$ in Q_E. The Rules 1 and 2 then give, after some calculation:

$$(9.8) \qquad V_E = Q_E^* + \left(\frac{1}{1 - \beta + \kappa}\right)\left[1 + \frac{\kappa(s^* - \beta/\sigma)}{\beta - \beta/\sigma}\right]\Delta_Y.$$

The corresponding function of productivity is of course $V_{Y/E} = V_Y - V_E$. Because the wage rate is in aggregate growth theories represented by $w = (1 - \beta)(Y/E)$, the symbol W will be sometimes in the following used for the productivity of labour Y/E.

Parameters:

It is a good methodological advice to keep the parameter values fixed as much as possible when applying to empirical data many-parameter models, such as necessarily appear in growth theory. Otherwise the accusation is near that one has used the available many parameters to justify "theoretical predictions" of whatever data. To avoid this criticism in advance a policy of fixed parameter values, following closely the 'traditional' estimates taken from or near those obtained by Denison (1961), has been obeyed in the applications of the present theory to business cycles, both in Chapters 8 and 9.

In the present chapter the parameter values will be fixed from the very beginning to be the same as they were in Chapter 8, but one obvious change is necessary here. It concerns the relevant balanced-growth output/capital ratio b^*, which of course now has the risen value $b^* = .50$ toward which the cycle center approaches asymptotically on the anomalous growth path (see Section 5.2). Hence we have now:

$$\underline{n} = .015, \quad \underline{\rho} = .02, \quad \underline{\beta} = .25 \quad \Longrightarrow \sup s^* = .1875, \quad \inf \sigma = 1.3333,$$
$$\underline{b^*} = .5, \quad \underline{s^*} = .13 \quad \Longrightarrow \sigma = 2.1, \quad \alpha^* = .06, \quad \omega = .1660786,$$
$$T = 2\pi/\omega = 37.832599, \quad \gamma = .0054761, \quad \underline{\kappa} = .417, \quad \underline{B} = .75.$$

The value .75 of the parameter B introduced in the formula (5.35) will be explained later. The length of period of a cycle, T, is indicated in theoretical time units and corresponds to 4 years in real time. This length of period in real time has been chosen in earlier applications too. The numerical value of κ has been again taken from Lucas (1988).

Integrations over the Period of a Cycle

To get correlations we have to compute integrals, over the period of a cycle, of products of powers of $(x - b^*), (s - s^*)$ and Δ, all of which are time functions. Products of the two first factors have already been taken care of: we have the solution (8.1)-(8.2) of the cycle equations, which makes integrals over time easy. As to the last factor we expand it for those integrals in an infinite series with exponential terms, i.e.

$$\Delta = b^* - b = b^* \sum_{r=1}^{\infty} (-1)^{r+1} B^r e^{-r\gamma t}. \tag{9.9}$$

The functional expressions and numerical values of the 'basis integrals' over a cycle, needed in the application to the anomalous cycles, can now be easily computed by using the formulae (8.1) and (8.2). In the following formulae the parameter a is as defined by (3.25), which can also be written as $a = (\beta - \beta/\sigma)^{-1}$. The parameter γ, of course, is as defined in (5.35). For the distance of the starting point of the cycle from the cycle center the notation

$$\delta = s(0) - s^*$$

will be used.

We need the following integrals over the cycle including the first and second degree of the variables $(s - s^*)$ and $(x - b^*)$ given in (8.1)-(8.2):

$$\frac{1}{T}\int_0^T (x - b^*)\,dt = \left(\frac{e^{\alpha^* T/2} - 1}{T}\right)\left(\frac{a}{1 - s^*}\right)\delta = (.489791)\,\delta,$$

$$\frac{1}{T}\int_0^T (s - s^*)\,dt = \left(\frac{e^{\alpha^* T/2} - 1}{T}\right)\left(\frac{4\alpha^*}{(\alpha^*)^2 + 4\omega^2}\right)\delta = (.11755)\,\delta,$$

$$\frac{1}{T}\int_0^T (x - b^*)^2\,dt = \left(\frac{e^{\alpha^* T} - 1}{T}\right) 2|D|^2 \left(\frac{1}{\alpha^*} - \frac{\alpha^*}{(\alpha^*)^2 + 4\omega^2}\right)\delta^2 = (4.194957)\,\delta^2,$$

$$\begin{aligned}\frac{1}{T}\int_0^T (s - s^*)^2\,dt &= \left(\frac{e^{\alpha^* T} - 1}{T}\right)\left[\frac{(\alpha^*)^2 + 4\omega^2}{8\alpha^*\omega^2} + \alpha^*\left(\frac{1.5 - (\alpha^*)^2/8\omega^2}{(\alpha^*)^2 + 4\omega^2}\right)\right]\delta^2 \\ &= (2.1533352)\,\delta^2,\end{aligned}$$

$$\frac{1}{T}\int_0^T (x - b^*)(s - s^*)\,dt = \left(\frac{e^{\alpha^* T} - 1}{T}\right)\left[\frac{a}{2(1 - s^*)}\right]\delta^2 = (1.0067913)\,\delta^2.$$

Then we have to calculate the integrals over the first cycle of the following functions, in which also the first or second degree of the change variable Δ appears:

$$\begin{aligned}\frac{1}{T}\int_0^T (x - b^*)\Delta\,dt &= \left(\frac{b^*}{T}\right) 2\omega|D|\sum_{r=1}^{\infty}(-1)^{r+1}B^r\left[\frac{e^{(\alpha^*/2 - r\gamma)T} - 1}{(\alpha^*/2 - r\gamma)^2 + \omega^2}\right]_o \\ &= (.0875566)\,\delta, \quad \text{where } \Sigma = 26.5, \\ \frac{1}{T}\int_0^T (s - s^*)\Delta\,dt &= \left(\frac{b^*}{T}\right)\sum_{r=1}^{\infty}(-1)^{r+1}B^r(\alpha^* - r\gamma)\,[\,\cdot\,]_o \\ &= (.0198241)\,\delta, \quad \text{where } \Sigma = 1.50, \\ \frac{1}{T}\int_0^T (s - s^*)^2\Delta^2\,dt &= S_1 + S_2 + S_3 = (.0827938)\,\delta^2.\end{aligned}$$

Here the infinite series S_1, S_2 and S_3 are as follows:

$$S_1 = \frac{2(b^*)^2|C|^2}{T}\sum_{r=2}^{\infty}(-1)^r B^r(r-1)\left[\frac{e^{(\alpha^*-r\gamma)T}-1}{\alpha^*-r\gamma}\right]_1 = (.0767655)\,\delta^2,$$

where $\Sigma = 22.5$,

$$S_2 = \frac{(b^*)^2\alpha^*}{T}\sum_{r=2}^{\infty}(-1)^r B^r(r-1)\left[\frac{e^{(\alpha^*-r\gamma)T}-1}{(\alpha^*-r\gamma)^2+4\omega^2}\right]_2 = (.004222512)\,\delta^2,$$

where $\Sigma = 10.65$,

$$S_3 = \frac{(b^*)^2(.5-(\alpha^*)^2/8\omega^2)}{T}\sum_{r=2}^{\infty}(-1)^r B^r(r-1)(\alpha^*-r\gamma)[\,\cdot\,]_2 = (.0018058)\,\delta^2,$$

where $\Sigma = .565$.

The remaining integrals are:

$$\begin{aligned}\frac{1}{T}\int_0^T (x-b^*)(s-s^*)\Delta\,dt &= \frac{b^*|D|}{T}\left\{\frac{\alpha^*}{2\omega}\left(\sum_{r=1}^{\infty}\right)_a(-1)^{r+1}B^r[\,\cdot\,]_1\right.\\ &\quad + \left.\left(\sum_{r=1}^{\infty}\right)_b(-1)^{r+1}B^r\left[2\omega-\frac{\alpha^*}{2\omega}(\alpha^*-r\gamma)\right][\,\cdot\,]_2\right\}\\ &= (.2043341)\,\delta^2, \quad \text{where } \Sigma_a = 62.9,\ \Sigma_b = 9.18,\end{aligned}$$

$$\begin{aligned}\frac{1}{T}\int_0^T (s-s^*)\Delta^2\,dt &= \frac{(b^*)^2}{T}\sum_{r=2}^{\infty}(-1)^r B^r(r-1)(\alpha^*-r\gamma)[\,\cdot\,]_o\\ &= (.0032643)\,\delta, \quad \text{where } \Sigma = .494,\end{aligned}$$

$$\frac{1}{T}\int_0^T \Delta\,dt = \frac{b^*}{\gamma T}\log\left(\frac{1+B}{1+Be^{-\gamma T}}\right) = (.2017471),$$

$$\begin{aligned}\frac{1}{T}\int_0^T \Delta^2\,dt &= \frac{(b^*)^2}{\gamma T}\left[\log\left(\frac{1+B}{1+Be^{-\gamma T}}\right)+\frac{1}{1+B}-\frac{1}{1+Be^{-\gamma T}}\right]\\ &= (.0407535).\end{aligned}$$

In the computation of integrals, where the integrand includes an infinite series, 20 first terms of the series were used, except for the two first such integrals in the above list, in which cases 12 first terms were considered to be sufficient.

9.2 Theoretical Explanation of Observed Anomalies in the Business Cycle

The general theory tells nothing specific concerning the period 1914-50, in which the anomalous phenomena were observed. To be specific we have to choose a model for this particular period, to be constructed on the basis of the general theory.

The model cycle can be found by studying the curve of the function $b(t)$ shown in Fig.3 (Section 5.2). In the period 1909-30 the average output/capital ratio represented by the parameter b was approximately a constant .30 in the U.S. economy. This is evident from Fig.3, based on the numbers in Table 1 given by Solow (1957). We can see that the parameter b then grew in the period 1930-48 from its original value .30 to the value .48. This growth obeyed a logistically rising curve toward an apparent asymptotic value .50 at $t = \infty$ of the function $b(t)$. Therefore we have to choose the constant B and the asymptotic value b^* in the function $b(t)$ of (5.35) as follows:

$$(9.10) \qquad B = .75 \quad \text{and} \quad b^* = .50 \Longrightarrow b(0) = .2857.$$

This locates the start of the anomalous cycles in the year 1930 in the middle of the Great Depression in the U.S.A., which according to Fig.3 started the anomalous growth.

We can also see from Fig.3 that the growth rate of the function $b(t)$ was largest at the beginning of the period 1930-50. Thus the first business cycle of that period can be expected to be the most anomalous cycle of this period. Therefore we shall study the *first anomalous cycle* as the model of the anomalous phenomena encountered in the U.S. (and also in the U.K.) business cycle in the period 1914-50.

Reduced Procyclicality of Consumption: U.S.A., U.K.

Using for the integrals the numerical values given above we first calculate, denoting the unknown initial state $s(0) - s^*$, as before, by δ, the expectation values

$$m^*(Y) \;=\; \beta\frac{1}{T}\int_0^T (x - b^*)\,dt = (.1224477)\,\delta,$$

$$m(\Delta_Y) = (\beta - \beta/\sigma)\frac{1}{T}\int_0^T \Delta dt = (.0264192)\,\delta,$$

then the variances and the mutual covariance of Q_Y^* and Δ_Y:

$$(\sigma_Y^*)^2 = \beta^2 \frac{1}{T}\int_0^T (x-b^*)^2\,dt - [m^*(Y)]^2 = (.2471914)\,\delta^2,$$

$$\sigma_{\Delta_Y}^2 = (\beta - \beta/\sigma)^2 \frac{1}{T}\int_0^T \Delta^2\,dt - [m(\Delta_Y)]^2 = .00000089,$$

$$\mathrm{cov}(Q_Y^*, \Delta_Y) = \beta(\beta - \beta/\sigma)\frac{1}{T}\int_0^T (x-b^*)\Delta\,dt - m^*(Y)m(\Delta_Y)$$
$$= (-.00036857)\,\delta.$$

Hence we get the expressions

$$\sigma_Y^2 = (\sigma_Y^*)^2 + \sigma_{\Delta_Y}^2 + 2\mathrm{cov}\,(Q_Y^*, \Delta_Y)$$
$$= \left(.2471914 + \frac{.0000008900}{\delta^2} - 2\frac{.00036857}{\delta}\right)\delta^2,$$

$$(\sigma_Y^*)^2 + \mathrm{cov}(Q_Y^*, \Delta_Y) = \left(.2471914 - \frac{.00036857}{\delta}\right)\delta^2.$$

What we want is the correlation

$$r_{CY} = \frac{\mathrm{cov}(C,Y)}{\sigma_C\sigma_Y} = \frac{(\sigma_Y^*)^2 + \mathrm{cov}(Q_Y^*, \Delta_Y)}{\sigma_Y^*\sigma_Y}.$$

Here we meet the first anomaly. While in ordinary business cycles the correlations over the detrended cycle were, at least to the second degree, entirely independent of the initial state $s(0)$ and thus of $\delta = s(0) - s^*$ (see Chapter 8), they now heavily depend on it. This is shown in Table 13, where the function $r_{CY}(\delta)$ has been calculated for some values of δ, with the parameter values as given in Section 8.1 above.

In Table 13 also the observed values of this correlation in the period 1914-50 in the U.S. and U.K. economies are given and comparted with the values of the theoretical function $r_{CY}(\delta)$. The values of δ, for which the theoretical value comes closest to the observed values in these two countries, are underlined.

TABLE 13. – THE FUNCTION $r_{CY}(\delta)$ OVER THE ANOMALOUS CYCLE COMPARED WITH THE OBSERVED U.S. AND U.K. VALUES.

δ	.000	001	.0011	.0015	.0020	.0025	.003	.005	.01
r_{CY}	−.78	−.38	−.32	.01	.40	.65	.79	.95	.99
U.S.	–	–	–	–	–	.51	–	–	–
U.K.	–	–	−.33	–	–	–	–	–	–

We can see from Table 13 that the fall in the procyclicality of consumption is the larger the smaller is the distance $\delta = s(0) - s^*$ from the fixed cycle center P_O. This distance indicates the point from which the anomalous cycles start. For the U.S. economy after the crash the result suggests a distance somewhere between .0020 and .0025. For the U.K. economy the corresponding distance is .0011.

The fall in procyclicality of consumption can be seen by comparing the ordinary U.S. value of the output correlation of consumption in Table 8, which is .85, to the observed anomalous value .51. It shows a fall of -.34. The fall in the corresponding theoretical value from Table 8, which was 1.00, to the anomalous theoretical value .65 is of the same order of magnitude, viz. -.35, which suggests the distance $\delta = .0025$ in the U.S. case.

Retained Procyclicality of Employment: U.S.A., U.K.

We first compute, using the numerical values of the basis integrals, the here needed magnitudes of the expectation value, the variance and the covariance with output of employment in the ordinary detrended cycles:

$$m_E^* = \left(\frac{\beta}{1-\beta+\kappa}\right)\left\{(1-s^*)\frac{1}{T}\int_0^T (x-b^*)\,dt - b^*\frac{1}{T}\int_0^T (s-s^*)\,dt\right\} = (.0786939)\,\delta,$$

$$(\sigma_E^*)^2 = \left(\frac{\beta}{1-\beta+\kappa}\right)^2\left\{(1-s^*)^2\frac{1}{T}\int_0^T (x-b^*)^2\,dt + \right.$$

$$- \quad 2(1-s^*)b^*\frac{1}{T}\int_0^T (s-s^*)(x-b^*)\,dt$$

$$+ \quad (b^*)^2 \frac{1}{T}\int_0^T (s-s^*)^2\,dt\Big\} - (m_E^*)^2 = (.1240301)\,\delta^2,$$

$$\mathrm{cov}(E,Y)^* = \left(\frac{\beta^2}{1-\beta+\kappa}\right)\Big\{(1-s^*)\frac{1}{T}\int_0^T (x-b^*)^2\,dt$$

$$- \quad b^*\frac{1}{T}\int_0^T (s-s^*)(x-b^*)\,dt\Big\} - m_E^* m_Y^* = (.1588632)\,\delta^2.$$

Here again the terms of higher than second order in δ have been omitted.

We need also the magnitude

$$\mathrm{cov}(Q_E^*,\Delta_Y) = \left(\frac{\beta}{1-\beta+\kappa}\right)\Big\{\left(\frac{1-s^*}{\beta}\right)\mathrm{cov}(Q_Y^*,\Delta_Y)$$

$$- \quad b^*\left[(\beta-\beta/\sigma)\frac{1}{T}\int_0^T (s-s^*)\Delta\,dt - m(s)m(\Delta_Y)\right]\Big\}$$

$$= (-.000220195)\,\delta.$$

This completes the preliminary work necessary for the calculation of the following functions of the anomalous business cycle:

$$\mathrm{cov}(E,Y) = \mathrm{cov}(E,Y)^* + \mathrm{cov}(Q_E^*,\Delta_Y) + \left(\frac{1}{1-\beta+\kappa}\right)\cdot$$

$$\cdot \left[1+\frac{\kappa(s^*-\beta/\sigma)}{\beta-\beta/\sigma}\right]\left[\sigma_{\Delta_Y}^2 + \mathrm{cov}(Q_Y^*,\Delta_Y)\right]$$

$$- \left[\frac{\beta}{(1-\beta+\kappa)^2}\right]\left[1+\frac{\kappa(s^*-\beta/\sigma)}{\beta-\beta/\sigma)}\right]\mathrm{cov}(sx,\Delta_Y)$$

$$= \left(.1588329 + \frac{.000000789237}{\delta^2} - \frac{.0005470367}{\delta}\right)\delta^2,$$

and

$$\sigma_E^2 = (\sigma_E^*)^2 + \left(\frac{1}{1-\beta+\kappa}\right)^2\left[1+\frac{\kappa(s^*-\beta/\sigma)}{\beta-\beta/\sigma}\right]^2\sigma_{\Delta_Y}^2$$

$$+ \left(\frac{2}{1-\beta+\kappa}\right)\left[1+\frac{\kappa(s^*-\beta/\sigma)}{\beta-\beta/\sigma}\right]\mathrm{cov}(Q_E^*,\Delta_Y)$$

$$- \left[\frac{2\beta}{(1-\beta+\kappa)^2}\right]\left[1+\frac{\kappa(s^*-\beta/\sigma)}{\beta-\beta/\sigma}\right]\text{cov}(sx,\Delta_Y)$$
$$= \left(.1239696+\frac{.00000069988}{\delta^2}-\frac{.0003905305}{\delta}\right)\delta^2.$$

Hence we can compute the correlations r_{EY} over the detrended anomalous business cycle as the function of the distance δ. The numerical values are listed in Table 14 together with the empirical values observed in the U.S. and the U.K. in the period 1914-50.

TABLE 14. – THE FUNCTION $r_{EY}(\delta)$ OVER THE DETRENDED ANOMALOUS BUSINESS CYCLE COMPARED WITH THE U.S. AND U.K. VALUES.

δ	.000	.0011	.0015	.002	.0025	.003	004	.005	.0075	.01
r_{EY}	1.00	.95	.89	.77	.76	.76	.78	.81	.85	.86
U.S.	–	–	–	–	.78	–	–	–	–	–
U.K.	–	.92	–	–	–	–	–	–	–	–

What strikes one when looking at these correlations, after having studied the corresponding correlations of consumption, is their continuously high values. Never do they fall below the value .76 obtained for $x = .0025$ and 003. This is in harmony with the empirical value .78 reported by Correia, Neves and Rebelo (1992) for the period 1914-50 in the U.S. economy. The observed U.S. value is close to the minimum level .76. or 77 of the theoretical correlations corresponding to the distances from .002 to .003.

The observed U.K. correlation is as high as .92, which again singles out the distance $\delta = .0011$.

Radically Reduced Procyclicality of Investment: U.S.A.

The form of the detrended cycle function V_I of investment, (6)(-7), gives immediately the variance and covariance we need:

$$\begin{aligned}\sigma_I^2 &= (F-G)^2(\sigma_Y^*)^2 + F^2\sigma_{\Delta_Y}^2 + F^4\sigma_{s\Delta}^2\\ &+ 2F(F-G)\text{cov}(Q_Y^*,\Delta_Y) - 2F^2(F-G)\text{cov}(Q_Y^*,s\Delta_Y)\\ &- 2F^3\text{cov}(s\Delta_Y,\Delta_Y) - 2F^3(\beta-\beta/\sigma)\text{cov}(sx,\Delta_Y),\end{aligned}$$

$$\begin{aligned}
\mathrm{cov}(I,Y) &= (F-G)\,(\sigma_Y^*)^2 + F\sigma_{\Delta_Y}^2 + (2F-G)\mathrm{cov}(Q_Y^*,\Delta_Y) \\
&- F^2\mathrm{cov}(Q_Y^*, s\Delta_Y) - F^2\mathrm{cov}(s\Delta_Y,\Delta_Y), \\
&- F^2(\beta-\beta/\sigma)\mathrm{cov}(sx,\Delta_Y).
\end{aligned}$$

In both expressions the terms of third and higher orders have been omitted, and the short notations $(s-s^*)\Delta_Y = s\Delta_Y$ and $(s-s^*)(x-b^*) = sx$ have been used.

The expectation value, variance and covariances, not encountered before but needed in these formulae, are easily computed from the numerical values of basis integrals given before:

$$\begin{aligned}
m(s\Delta_Y) &= (\beta-\beta/\sigma)\frac{1}{T}\int_0^T (s-s^*)\Delta\, dt \\
&= (.00259600)\,\delta, \\
\sigma_{s\Delta_Y}^2 &= (\beta-\beta/\sigma)^2\frac{1}{T}\int_0^T (s-s^*)^2\Delta^2\, dt - [m(s\Delta_Y)]^2 \\
&= (.001412961)\,\delta^2,
\end{aligned}$$

$$\begin{aligned}
\mathrm{cov}(Q_Y^*, s\Delta_Y) &= \beta(\beta-\beta/\sigma)\frac{1}{T}\int_0^T (x-b^*)(s-s^*)\Delta\, dt - m^*(Y)m(s\Delta_Y) \\
&= (.0063716)\,\delta^2, \\
\mathrm{cov}(s\Delta_Y,\Delta_Y) &= (\beta-\beta/\sigma)^2\frac{1}{T}\int_0^T (s-s^*)\Delta^2\, dt - m(s\Delta_Y)m(\Delta_Y) \\
&= (-.000012607)\,\delta, \\
\mathrm{cov}(sx,\Delta_Y) &= (\beta-\beta/\sigma)\frac{1}{T}\int_0^T (s-s^*)(x-b^*)\Delta\, dt - m(sx)m(\Delta_Y) \\
&= (.000159400)\,\delta^2.
\end{aligned}$$

By combining the above results we get the following numerical expressions:

$$\sigma_I^2 = \left(6.548721 + \frac{.000047336}{\delta^2} - \frac{.01407100}{\delta}\right)\delta^2,$$

$$\mathrm{cov}(I,Y) = \left(.7354666 + \frac{.0000061538}{\delta^2} - \frac{.0037506}{\delta}\right)\delta^2.$$

This gives for the correlation of investment with output over a detrended cycle, $r_{I,Y} = \mathrm{cov}(I, Y)/\sigma_Y \sigma_I$, as a function of δ the numerical values listed in Table 15 together with the empirical value observed in the U.S. in the period 1914-50.

TABLE 15. – THE FUNCTION $r_{IY}(\delta)$ OVER THE DETRENDED ANOMALOUS BUSINESS CYCLE COMPARED WITH THE U.S. VALUE.

δ	.000	.001	.0015	.0020	.0023	<u>.0025</u>	<u>.0030</u>	<u>0040</u>	.0050	.0075	.01
r_{IY}	1.00	.78	.58	.37	.28	.24	.20	.21	.26	.36	.41
U.S.	–	–	–	–	–	–	.16	–	–	–	–

Again we can see the fall in procyclicality depending heavily on the distance δ of the starting point of the anomalous cycles from the fixed cycle center P_O. The U.S. value .16 is close to the three underlined values of δ, one of which is again the number .0025 encountered in the cases of consumption and employment as well.

It is obvious from Tables 13-15 that the value $\delta = .0025$ gives the best theoretical predictions for the U.S. economy in an overall comparison of the predictions and empirical results in those tables. The summary of the results concerning the anomalous cycle in the U.S.A. is given in Table 16. Let it be recalled that the empirical numbers come from the work of Correia, Neves and Rebelo (1992). Their analysis did not include productivity in the U.S. case, which is why productivity does not appear in Table 16 either.

TABLE 16. – THE COMPARISON OF MODEL CORRELATIONS WITH THE EMPIRICAL ONES: U.S.A.

Variable	*Model* (0025)	*TheU.S. economy*
Consumption	.65	.51
Investment	.24	.16
Employment	.76	.78

The errors of prediction are .14, .08 and .02, and the sum of error squares is .0264. This is in fact smaller than the sum of error squares over the same three variables in Table 8, which is .0514. Thus the above predictions of correlations over the anomalous cycle in the U.S. economy are at least on the same level of accuracy than were the predictions of the corresponding correlations over the ordinary cycle, which on the other hand were much more accurate than are the predictions of any of the conventional stochastic models. All this testifies both of the existence of anomalous business cycles – which has sometimes been doubted – and of the realism of their theoretical approximation by means of the minimal macroeconomic dynamics constructed in Chapter 3 of this book.

Another interesting consequence from Table 16 concerns the value $\delta = s(0) - s^* = .0025$, which indicates that the starting point $S_A = (s(0), b^*)$ of anomalous cycles was at the very small distance .0025 from the fixed cycle center $P = (s^*, b^*)$ of the ordinary cycles. In other words, the anomalous cycles started from a collapse of the ordinary cycles. The turning point S_A is unobservable in economic statistics, because the anomalous cycles continued the cycles immediately after the collapse of the ordinary ones.

These consequences are made from a relatively small number of numerical facts and are therefore best to be regarded as tentative. On the other hand these numerical facts belong to what is considered to be essential quantitative indexes telling about the business cycle.

Productivity Turned Anticyclical: U.K.

By using the formulae (here the notation $Y/E = W$ is used)

$$\begin{aligned} \sigma_W^2 &= \sigma_Y^2 + \sigma_E^2 - 2\text{cov}(E, Y) \text{ and} \\ \text{cov}(W, Y) &= \sigma_Y^2 - \text{cov}(E, Y) \end{aligned}$$

we get for the function $r_{WY}(\delta)$ the values indicated in Table 17.

TABLE 17. – THE FUNCTION $r_{WY}(\delta)$ OVER THE DETRENDED ANOMALOUS BUSINESS CYCLE COMPARED WITH THE EMPIRICAL U.K. VALUE.

δ	.000	.0005	<u>.0011</u>	.0015	.002	.0025	.003	.005	.01
r_{WY}	1.00	.40	−.01	.08	.32	.45	.55	.68	.73
U.K.	–	–	−.61	–	–	–	–	–	–

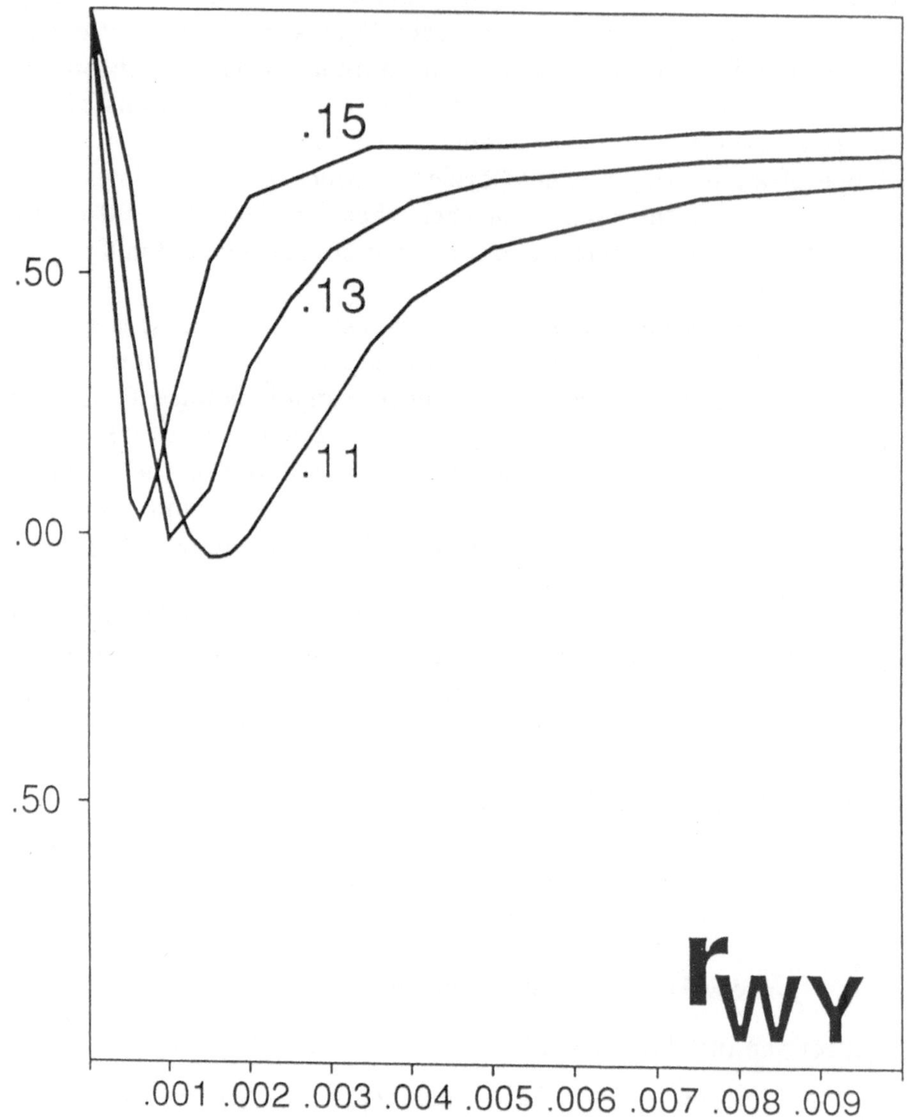

FIGURE 12.
AN ILLUSTRATION OF THE DEPENDENCE OF THE ANOMALOUS OUTPUT CORRELATION OF THE PRODUCTIVITY OF LABOUR W ON THE SAVINGS RATE PARAMETER.

Again, in Table 17, the value of δ that gives the theoretical correlation closest to the empirical one has been underlined, and again it is .0011 for the U.K. economy. But the quantitative error in the prediction of this correlation, as shown shown by Table 17, is enormous: .60. When compared with the very small errors, .01 and .03, shown by the earlier U.K.-predictions of the output correlation of consumption and of the output correlation of employment in Tables 13 and 14, respectively, only one conclusion is possible: while the theory of anomalous cycles gives very good quantitative predictions of the latter two correlations, there can be no talk of a similar success in the case of the correlation of productivity.

On the other hand Fig.12 shows that there is a characteristic plunge in the value of the function $r_{WY}(\delta)$ for small values of δ, and that such a plunge is there for all the parameter values $s^* = .11, 13$ and .15. The plunge of the curve for the parameter value $s^* = .13$ used in these applications takes place at the distance $\delta = .0011$. We may call this a qualitative explanation of the empirical value $r_{WY} = -.61$: the plunge is there at the correct distance from the fixed cycle center P but is not deep enough.

As can be seen from Table 18, theoretical prediction was even worse in the case of the output correlation of investment in the U.K. (period 1914-50).

TABLE 18. – THE COMPARISON OF MODEL CORRELATIONS WITH THE EMPIRICAL ONES: U.K.

Variable	*Model* (0011)	*TheU.K. economy*
Consumption	−.32	−.33
Employment	.95	.92
Productivity	−.01	−.61
Investment	.78	−.41

Table 18 also tells that there is, in the case of the U.K. economy,

1) a perfect agreement of theory and data as far as the anomalous correlations with output of consumption and employment are concerned,

2) a perfect agreement about the model distance $\delta = .0011$ in the cases of consumption, investment and productivity, but

3) only a qualitative explanation is given by this theory of the fall in the output correlation of productivity, and

4) a perfect failure of the present theory to cope with the anomalous output correlation of investment.

Obviously the anticyclicality of both productivity and investment in the U.K. need another explanation, e.g. the governmental regulation. But the anomalous cycles also in the case of the U.K. started from the collapse of ordinary cycles at a very short distance from the fixed cycle center P: again the turning point is unobservable in economic statistics, because the anomalous cycles continue immediately after the collapse of the ordinary ones.

Causal Background

The present theory of anomalous cycles predicted rather well the anomalous output correlations observed in the period 1914-50 in the U.S. economy, and at least qualitatively it predicted also 75 % of observed correlations in the case of the U.K. economy. We may ask whether the dynamics of anomalous cycles as given by this theory can be explained in any less formal way than by the mathematical formulae constructed in the present Chapter 9?

One can roughly illustrate the cause of the anomalous correlations, embedded in the formulae, by a visual image. The ultimate cause, in terms of the present theory of the anomalous cycles, of course is the moving cycle center P_A of the anomalous growth type (cf. Section 5.2). The cycle center rises along a straight line, viz. the line $s = s^*$ on the plane (s, x), where the cycles are originated. The cycles "leave behind" in this movement of their center, and thus produce a phenomenon of retardation: the correlations fall. The retardation depends on the magnitude of the component of the cycle function in the direction of the x-axis, i.e. in the direction of the movement of the cycle center. But this is only a rough picture of what happens, since the cycles too are deformed in the process.

Appendix (Chapter 9)

The Dependence of the Predicted Anomalous Correlations on Savings Rate

TABLE 19.
THE PREDICTED ANOMALOUS OUTPUT CORRELATIONS OF CONSUMPTION AND INVESTMENT.[1]

δ	r_{CY}			r_{IY}		
	.11	.13	.15	.11	.13	.15
.0000	−.784	−.786	−.785	1.000	1.000	1.000
.0005	−.701		−.456			
.0006	−.680		−.340		+.886	+.911
.0010	−.575	−.386		+.946	+.786	+.654
.0011		−.316				
$.0012_5$	−.490		+.527			+.486
.0013	−.471		+.571			+.460
.0014			+.641			+.417
.0015	−.389	+.008	+.706		+.584	+.384
.0020	−.141	+.398	+.868	+.754	+.372	+.338
.0023					+.284	
.0025		+.652	+.929		+.245	+.375
.0030	+.382	+.789	+.956	+.476	+.200	+.425
.0033	+.501					
.0035	+.572			+.369		
.0040	+.699		+.979	+.298	+.216	+.507
.0050	+.838	+.948	+.987	+.230	+.265	+.560
.0060	+.903					
.0075	+.947		+.995	+.219	+.362	+.630
.0100	+.974	+.990	1.000	+.247	+.417	+.662

[1]Only the $C-$ and $I-$points marked in Figures 13 and 15, respectively, are indicated in Table 19. The numbers .11,.13 and .15 are the chosen values for the savings rate parameter s^*.

TABLE 20.
THE PREDICTED ANOMALOUS OUTPUT CORRELATIONS OF EMPLOYMENT AND PRODUCTIVITY.[2]

δ	r_{EY}			r_{WY}		
	.11	.13	.15	.11	.13	.15
.0000	1.000	1.000	1.000	1.000	1.000	1.000
.0005	+.997	+.994	+.978	+.675	+.404	+.067
.0006			+.963			+.025
.0008				+.278		+.085
.0010	+.986	+.963	+.863	+.108	−.009	+.229
.0011		+.952				
.0012		+.939				
$.0012_5$			+.810	−.003		
.0013			+.804			
.0014			+.795			
.0015		+.891	+.791	−.043	+.086	+.523
$.0017_5$	+.938			−.037		
.0020	+.912	+.769	+.806	−.000	+.323	+.645
.0025	+.850	+.763	+.831		+.451	
.0030		+.757			+.547	
.0035	+.750		+.864	+.370		+.743
.0040	+.732	+.783	+.874	+.454	+.639	
.0045	+.728					
.0050	+.733	+.810	+.887	+.555	+.680	+.744
.0075	+.779	+.848	+.894	+.649	+.720	+.775
.0100	+.810	+.865	+.910	+.681	+.735	+.789

[2]Only the W- and E−points marked in Figures 12 and 14, respectively, are indicated in Table 20. The numbers .11,.13 and .15 are the chosen values for the savings rate parameter s^*.

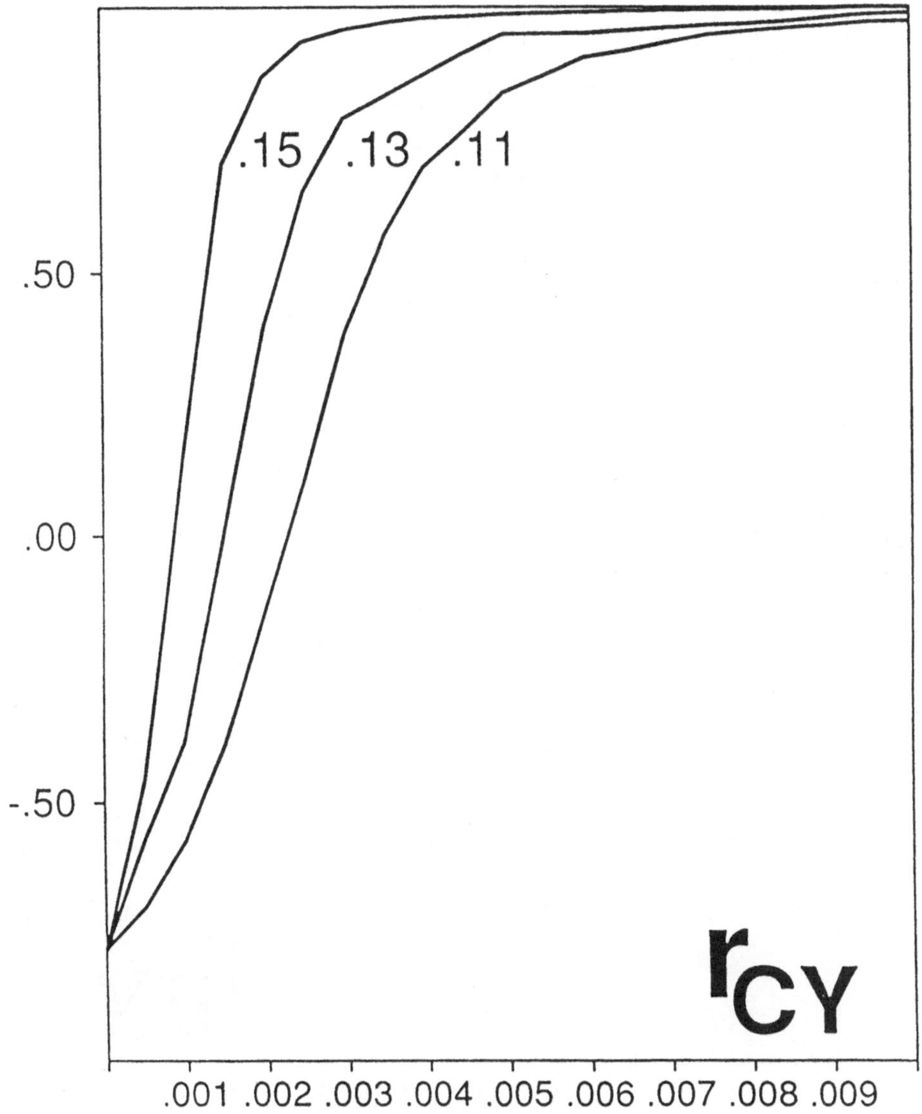

FIGURE 13.
AN ILLUSTRATION OF THE DEPENDENCE OF THE ANOMALOUS CORRELATION OF CONSUMPTION ON THE SAVINGS RATE PARAMETER.

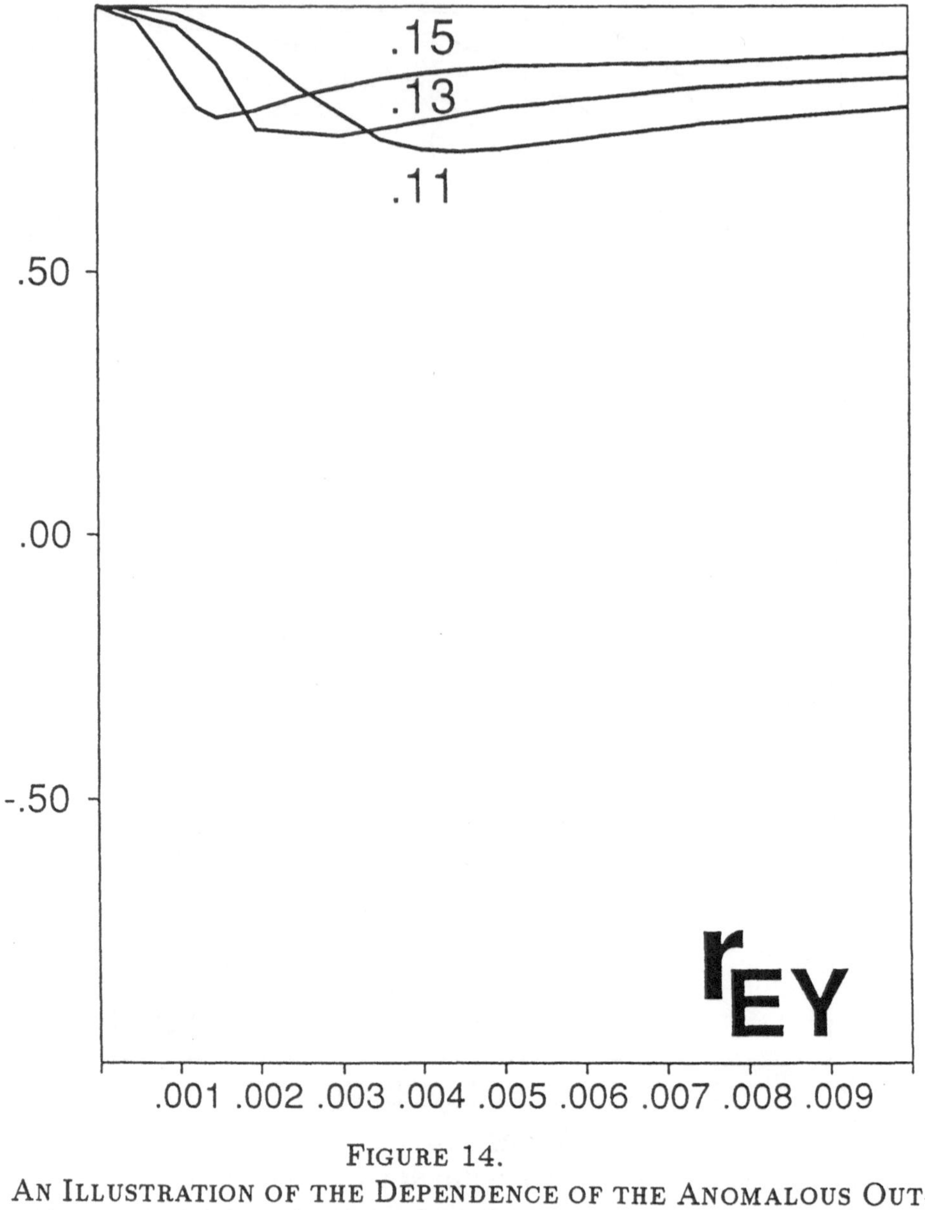

FIGURE 14.
AN ILLUSTRATION OF THE DEPENDENCE OF THE ANOMALOUS OUTPUT CORRELATION OF EMPLOYMENT ON THE SAVINGS RATE PARAMETER.

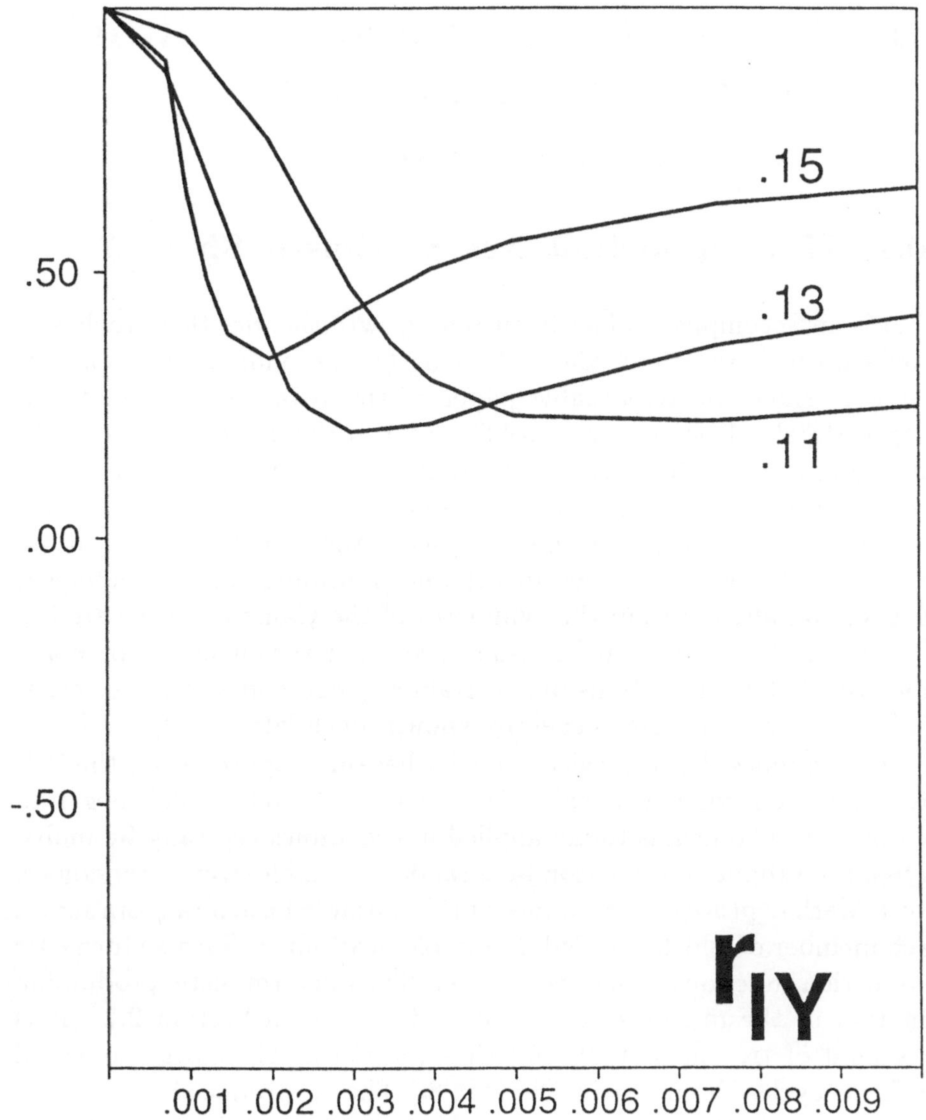

FIGURE 15.
AN ILLUSTRATION OF THE DEPENDENCE OF THE ANOMALOUS OUTPUT CORRELATION OF INVESTMENT ON THE SAVINGS RATE PARAMETER.

10 The Role of Stochastic Shocks in the Business Cycle

(*Complementary Evidence*)

10.1 How Important Are Stochastic Shocks?

It is an accomplished fact in current growth theories that stochastic shocks do not essentially affect the long-term economic development. But ever since the remarkable paper of the Russian mathematician Eugen Slutsky from the year 1927 was translated and published in English in a completed form (Slutsky, 1937), the economists have been fascinated by the idea that the business cycles may be produced in the way discussed by Slutsky in his analysis of the Gaussian distribution. He showed that random series are capable of forming cyclic phenomena. This follows already from the symmetry of the Gaussian curve around the mean of the series, but he also showed that the effect can be made more visible by summations over certain sequences in a random series. These discoveries became generally known much later.

In economics the findings of Slutsky became popular when the failure of the then existing causal (and reductionist) cycle models began to be obvious. The idea is today applied in economics typically by multiplying the production function by a random variable, which undergoes, say, a Markov process. The values of this variable form a random series, each member of which is called a technological shock. Such a stochastic production function is then treated just like a deterministic production function in the maximization process described in Section 2.2. It is this kind of treatment that underlies the stochastic shock models of business cycles discussed in Chapter 8. The so produced oscillations of output and other economic variables around a trend, often a loglinear one, are then considered as models of the business cycles, with the results quoted in Chapter 8.

An explanation of economic phenomena on the basis of stochastic shocks may be questioned as a matter of principle, because it is really not a causal explanation, which has been the usual characteristic of scientific explanation so far. This criticism was expressed by

Richard Day (1992) in the passage quoted on p.62 of this book. The main criticism of the conventional method based on the superposition of stochastic shocks on microeconomics comes however from the fact that such models are in this book shown to be incapable of explaining the basic macroeconomic facts 7, 10-12, or 13-15 (see Chapters 6-9) – not to speak of the anomalous phenomena 16-17.

What also seems questionable to the present author is that in the method of stochastic optimization applied in those models the economic agents are supposed to be capable of reacting rationally to the irregular shocks: this idea is of course involved in the maximization of utilities which already include the stochastic shock variables. So far as the business cycles are thought to be produced mainly by the shocks, no other method of constructing the cycles are available. However, isn't it too much to expect that all the economic agents are capable of reacting rationally to the irregular shocks? Wouldn't it be more natural to assume that most agents react to the trends, observable over a certain interval of time, and not to the irregular shocks?

If mathematical macroeconomics is based on the theory constructed in Chapter 3 , we can easily realize the latter, more realistic situation. We already have the "cycles". What we need is to complete the theory by superposing the technological (or other) shocks upon the causal cycles as studied in Chapter 8 (or, for anomalous cycles, in Chapter 9). This is a problem different from that met with in the usual stochastic models of the business cycle.

It will be shown in this chapter that when superposed on the minimal macroeconomic dynamics constructed in Chapter 3 the shocks have a minor function: they appear only as perturbations of the nonstochastic causal cycles. How much do such shocks affect the correlations and standard deviations of economic variables over detrended cycles: this is the problem that will be studied in this last chapter. Technological shocks of the usually applied size will be superposed upon the ordinary cycles of Chapter 8 in order to see how great is their influence on correlations and standard deviations.

10.2 Construction of Stochastic Cycle Functions

Random Series with a Definite Mean and SD

We can construct a random series in many ways. Here the following method is used. Take the seven first decimals of two irrational numbers, say $\pi - 3$ and $\sqrt{3} - 1$. Multiply each of the so obtained two numbers from the interval [0,1] by a prime number, say 317. Take only the decimal part of each product. Multiply them again by 317 and take the decimal parts. Go on until you have in both series 305 numbers belonging to the interval [0,1]. We can name them: let they represent the two random variables u_1 and u_2, with a constant distribution in the interval [0,1].

Hence we can proceed by constructing the new random variables

$$\begin{aligned} x_1 &= \sqrt{2\log(1/u_1)}\,\sin 2\pi\, u_2, \\ x_2 &= \sqrt{2\log(1/u_1)}\,\cos 2\pi\, u_2. \end{aligned}$$

The number series constructed in this way give two variables with the normal distribution $\mathbf{N}(0,1)$. Random variables having a normal distribution with any desired means μ_i and standard deviations σ_i are then given by

$$z_1 = \mu_1 + \sigma_1 x_1, \quad z_2 = \mu_2 + \sigma_2 x_2.$$

This is a quick method of getting two random series with given means and standard deviations. Here we shall need only one of them, say z_1. The random series u_i and z_i, each of them including 305 numbers, are given in the Appendix to this chapter. Also the value of the ordinary cycle function $Q_Y = \beta(x - b^*)$ at the points of time $t_n = (.2)n$ for $n = 0, 1, 2, ..., 304$ are given there. These times correspond to an approximate division of the full cycle of the function $x - b^*$ in equal intervals with the length .02 in terms of the theoretical time unit [TU].

Stochastic Cycle Functions and Their Effects on Correlations and Standard Deviations

How the technological shocks affect each economic variable? It is usual in current shock models of business cycles to assume a technological shock variable with a lognormal distribution having the mean one. The multiplication of a given output Y by such a stochastic factor is equivalent to adding to the growth rate of Y a random variable of the type z, with the mean zero. Let us call it z_Y and identify it with the variable z_1 constructed above. This gives

$$z_Y = \sigma_1 x_1, \tag{10.1}$$

with a standard deviation σ_1 to be chosen later.

To consider the effects of technological shocks on the consumption C and the investment I we have to study the expressions indicating the relations of their growth rates with that of output:

$$\frac{\dot{C}}{C} = \frac{\frac{d}{dt}(1-s)}{1-s} + \frac{\dot{Y}}{Y}, \quad \frac{\dot{I}}{I} = \frac{\dot{s}}{s} + \frac{\dot{Y}}{Y}.$$

The shock variables of these economic variables must obey the same equations:

$$z_C = z_{1-s} + z_Y, \qquad z_I = z_s + z_Y.$$

With the approximation

$$\frac{\dot{s}}{s} \equiv -\frac{1-s}{s}\left(\frac{\frac{d}{dt}(1-s)}{1-s}\right) \approx -\frac{1-s^*}{s^*}\left(\frac{\frac{d}{dt}(1-s)}{1-s}\right)$$

and assuming no consumption shocks, $z_C = 0$, this gives:

$$z_s = -\left(\frac{1-s^*}{s^*}\right) z_{1-s}.$$

Accordingly we have:

$$z_C = 0, \quad z_I = \frac{1-s^*}{s^*} z_Y + z_Y = \frac{\sigma_1}{s^*} x_1. \tag{10.2}$$

To find the effects of shocks upon the employment E and the productivity of labour $Y/E = W$, we have to discuss the production function (3.63) when written for the growth rates. In view of (3.19) this reads (for the ordinary cycles around a balanced-growth path on which $\psi = \alpha^* =$ Constant):

$$\frac{\dot{Y}}{Y} = \beta\frac{\dot{K}}{K} + (1-\beta+\kappa)\frac{\dot{h}}{h} - (1-\beta)m.$$

It follows that we have to put:

$$z_Y = \beta z_K + (1-\beta+\kappa)z_h - (1-\beta)m.$$

Here we have to decide to which extent we shall now assume the technological shocks to be caused by shocks in physical capital (e.g. new oil fields found, or old ones closed, or oil price changed) and to which extent we suppose they are due to shocks in human capital (e.g. new methods of production discovered). For the sake of simplicity we can make the assumption that

$$(10.3) \qquad z_K = 0, \quad \text{thus } z_h = \frac{\sigma_1 x_1}{1-\beta+\kappa} + \frac{(1-\beta)m}{1-\beta+\kappa}.$$

This gives, since $E = h\alpha^*/k$ and $W = Y/E$:

$$(10.4) \qquad z_E = \frac{\sigma_1 x_1}{1-\beta+\kappa} - \frac{\kappa m}{1-\beta+\kappa}, \quad z_W = z_Y - z_E.$$

The stochastic shock variables (1)-(4) have to be added to the corresponding cycle functions Q_X. This gives the stochastic economic variables we need:

$$(10.5) \quad Y_{st} = Q_Y + \sigma_1 x_1, \; C_{st} = Q_C = \frac{1}{\sigma}Q_Y, \; I_{st} = Q_I + \frac{\sigma_1 x_1}{s^*},$$

$$(10.6) \quad E_{st} = Q_E + \frac{\sigma_1 x_1}{1-\beta+\kappa} - \frac{\kappa m}{1-\beta+\kappa}, W_{st} = Y_{st} - E_{st}.$$

We can call these functions X_{st} the *stochastic cycle functions.* Note that in the expression for C_{st} the symbol σ of course means the risk aversion coefficient, not a standard deviation (to make the distinction clear the latter are always equipped with a subscript).

By inserting the linearized ordinary cycle functions defined by the equations (8.7)-(8.9) in the stochastic cycle functions (5)-(6) we can easily deduct the following formulae for the variances of these stochastic variables over a detrended cycle:

$$(10.7) \quad \sigma^2_{Y_{st}} = \sigma^2_Y + 2\sigma_1 \mathrm{cov}(x_1, Q_Y) + \sigma^2_1,$$

$$(10.8) \quad \sigma^2_{C_{st}} = \sigma^2_C = \left(\frac{\sigma_Y}{\sigma}\right)^2,$$

$$\begin{aligned}
(10.9)\quad \sigma^2_{I_{st}} &= \sigma^2_I + (2B\sigma_1/s^*)\,\mathrm{cov}(x_1, Q_Y) + (\sigma_1/s^*)^2, \\
&\quad \text{with } B = \sigma_I/\sigma_Y; \\
(10.10)\quad \sigma^2_{E_{st}} &= \sigma^2_E + \left[\frac{2\sigma_1(1-s^*)}{(1-\beta+\kappa)^2}\right]\mathrm{cov}(x_1, Q_Y) \\
&\quad - \left(\frac{2\sigma_1\beta b^*}{1-\beta+\kappa}\right)\mathrm{cov}(x_1, s-s^*) + \left(\frac{\sigma_1}{1-\beta+\kappa}\right)^2, \\
(10.11)\quad \sigma^2_{W_{st}} &= \sigma^2_{Y_{st}} + \sigma^2_{E_{st}} - 2\,\mathrm{cov}(E_{st}, Y_{st}).
\end{aligned}$$

For their covariances with output we get likewise:

$$\begin{aligned}
(10.12)\quad \mathrm{cov}(I_{st}, Y_{st}) &= \mathrm{cov}(I, Y) + (B + 1/s^*)\sigma_1\,\mathrm{cov}(x_1, Q_Y) \\
&\quad + \sigma_1^2/s^*, \\
(10.13)\quad \mathrm{cov}(E_{st}, Y_{st}) &= \mathrm{cov}(E, Y) + \left(\frac{\sigma_1(2-s^*)}{1-\beta+\kappa}\right)\mathrm{cov}(x_1, Q_Y) \\
&\quad - \left(\frac{\sigma_1\beta b^*}{1-\beta+\kappa}\right)\mathrm{cov}(x_1, s-s^*) + \frac{\sigma_1^2}{1-\beta+\kappa}, \\
(10.14)\quad \mathrm{cov}(W_{st}, Y_{st}) &= \sigma^2_{Y_{st}} - \mathrm{cov}(E_{st}, Y_{st}).
\end{aligned}$$

For the consumption C, to which no shock variable is attached, we get directly the correlation:

$$(10.15)\qquad r_{C_{st},Y_{st}} = \frac{\sigma_Y}{\sigma_{Y_{st}}} + \sigma_1\frac{\mathrm{cov}(x_1, Q_Y)}{\sigma_Y\sigma_{Y_{st}}}.$$

10.3 Numerical Calculations and Conclusions

(i). The first term in each of the equations (7)-(10) and (12)-(13) is the variance or covariance, respectively, of the corresponding non-stochastic variable in the ordinary business cycles. We can take their numerical values directly from the calculations made in Chapter 8.

(ii). In the same way we get the numerical values of the constants

$$\sigma = 2.2916666, \quad B = 4.7702278, \quad s^* = .13, \quad 1 - \beta + \kappa = 1.167,$$
$$\beta = .25, \quad b^* = .3.$$

(iii). The computation of $\mathrm{cov}(x_1, Q_Y)$. Here we must first of course construct the random series x_1 (see Appendix to this chapter) as well as the corresponding series of the values of the cycle function $Q_Y = \beta(w - b^*)$. The latter can be constructed on the basis of the linearized cycle equation (3.47), which gives:

$$(10.16) \qquad Q_Y = -e^{\alpha^* t/2} \left[\frac{\beta a(\alpha^2 + 4\omega^2)}{4(1 - s^*)\omega} \right] (\sin \omega t) x_o.$$

Here the unknown initial state $s(0) - s^*$ has been denoted by x_o, and the constants, again obtained directly from Section 21, have the values:

$$\alpha^* = .036, \quad \beta = .25, \quad \omega = .1034851, \quad a = 7.0967745.$$

To match the values of Q_Y with the series x_1 the length $T = 2\pi/\omega$ of a period of its cycle (in theoretical time units) must be divided in 304 equal intervals. Since $T = 60.715844$ this gives the following series of time points:

$$(10.17) \qquad t_n = (.2)n \quad \text{with} \quad n = 0, 1, 2, ..., 304.$$

The values of Q_Y at these points of time are easily computed by means of the formula obtained by inserting (17) and the above values of constants in the equation (16), which gives:

$$(10.18) \quad Q_Y(n) = (-.217421)\, e^{(.0036)n} \sin(1.1858519n)^\circ x_o.$$

Here the angle has been already transformed to degrees. Both series $Q_Y(n)$ and $x_1(n)$ are given in the Appendix. The series of $x_1(n)$ is long enough to bring the mean very close to its expectation value zero, and we can compute the covariance by using the formula

$$(10.19)\quad \operatorname{cov}(x_1, Q_Y) = \frac{1}{305}\sum_{n=0}^{304} x_1(n)Q_Y(n) = -(.0242242)\,x_o.$$

(iv). The computation of $\operatorname{cov}(x_1, s - s^*)$. Here we can shorten the calculation by using the formula

$$(10.20)\qquad s - s^* = \left(\frac{\alpha^*}{2\beta\omega A}\right) Q_Y + Rx_o,$$

$$A = .8696843, \quad R = e^{\alpha^* t/2}\cos\omega t,$$

obtained from (3.46) and (3.47). Appendix at the end of this Chapter 10 gives also the series $R(n)$ for $n = 0, 1, 2, ..., 304$ needed in this computation. We compute first:

$$(10.21)\quad \operatorname{cov}(x_1, Rx_o) = \frac{1}{305}\sum_{n=0}^{304} x_1(n)R(n)x_o = (-.0107423)\,x_o.$$

Combining the results (19)-(21) we then get:

$$(10.22)\qquad \operatorname{cov}(x_1, s - s^*) = (-.0301217)\,x_o.$$

Calibration

To tell what there is to be calibrated we need to know how far can take us the numerical knowledge we already have. After the above preliminary calculations we can give the numerical expressions, where only two unknown constants appear, viz. the standard deviation σ_1 of the random variable z_1 and the initial state x_o of the cycle function $Q_Y(s, x)$:

$$(10.23)\ \sigma^2_{Y_{st}} = (.0784541)\,x_o^2 - (.0484484)\,\sigma_1 x_o + \sigma_1^2,$$

$$(10.24)\ \frac{\sigma_C}{\sigma_Y} = (6.944552)\,x_o, \quad r_{C_{st},Y_{st}} = (15.914585)\,x_o$$
$$-(4.9139279)\,\sigma_1,$$

$$(10.25)\ \sigma^2_{I_{st}} = (1.7865789)\,x_o^2 - (1.7784393)\,\sigma_1 x_o + (59.171597)\,\sigma_1^2,$$

$$(10.26)\ \sigma^2_{E_{st}} = (.0398651)\,x_o^2 - (.027632)\,\sigma_1 x_o + (.7342742)\,\sigma_1^2,$$

$$(10.27)\ \sigma^2_{W_{st}} = (.017479)\,x_o^2 - (.0023186)\,\sigma_1 x_o + (.8773762)\,\sigma_1^2.$$

The remaining expressions of covariances are:

$$\text{(10.28)} \quad \operatorname{cov}(I_{st}, Y_{st}) = (.3743856)\, x_o^2 - (.3019385)\, \sigma_1 x_o + (7.6923076)\, \sigma_1^2,$$

$$\text{(10.29)} \quad \operatorname{cov}(E_{st}, Y_{st}) = (.0504201)\, x_o^2 - (.0368809)\, \sigma_1 x_o + (.856898)\, \sigma_1^2,$$

$$\text{(10.30)} \quad \operatorname{cov}(W_{st}, Y_{st}) = (.028034)\, x_o^2 - (.0115675)\, \sigma_1 x_o + (.143102)\, \sigma_1^2.$$

To calibrate as close to empirical values as possible, we first of course choose the standard deviation of the stochastic output variable Y_{st} to be equal to the observed standard deviation, which in the U.S. economy used as empirical comparison in the stochastic optimization models is .0176. The same calibration has been used in the stochastic optimization models of Kydland and Prescott, Hansen and Rogerson,and Danthine and Donaldson.

According to Hansen (1985, p.320), " A data analysis suggests that [the standard deviation of the technological shock] could reasonably be expected to lie in the interval [.007,.010]." To keep as close as possible to the choices made in the stochastic optimization models, with whose results the outcome of these calculations will be compared here, the Hansen-Rogerson value $\sigma_1 = .00712$ is chosen for the constant σ_1. Thus the technological shocks here applied are of the same order of magnitude as are the Hansen shocks (they are not exactly the same because of the logarithm function involved).

Thus we have the following calibration:

$$\text{(10.31)} \quad \sigma_{Y_{st}} = .0176, \quad \sigma_1 = .00712.$$

These values give, in view of (17):

$$\text{(10.32)} \quad x_o = .0597035.$$

Small But Not Negligible Effects of Shocks

After the above calibration we can first find out which proportions of the standard deviation of output are explained by the nonstochastic and stochastic approaches, respectively, in the dynamics of this book. We get:

$$(10.33) \qquad \sigma_Y = (.2800967)\,x_o = .0167227 = 95\% \text{ of } .0176,$$

leaving only 5% of the empirical value .0176 to be explained by the technological shocks. This corresponds to 10% of the variance of output.

By inserting the values (31) and (32) in the expressions (24)-(30) we get the other numerical predictions given by the stochastic approach. They are compared in Table 21 with the numerical predictions of the nonstochatic approach as given in Chapter 8. We can see that the shocks have contributed very little to the predictions of the nonstochastic approach of Chapter 8, with the exception perhaps of the standard deviation of investment. But even the change in this case is in terms of precentages only 10%.

The sums of error squares in Table 22 show that both versions, with or without shocks, of the dynamics of constructed in this book do far better than the Danthine-Donaldson,Hansen-Rogerson or Kydland-Prescott models in standard deviations and clearly better also in correlations.

The lines describing the predictions of the two versions of this theory in Figs. 16 and 17 follow – at a certain distance – the empirical line, those of the stochastic RBC models together stray off that line.

TABLE 21. STANDARD DEVIATIONS AND CORRRELATIONS OVER A CYCLE PREDICTED BY THE NONSTOCHASTIC (NON-ST) AND STOCHASTIC (STOCH) VERSIONS OF THIS THEORY.

	Standard deviation proportions :			*Correlations with output :*		
Variable	*Non – st*	*Stoch*	Δ%	*Non – st*	*Stoch*	Δ%
Output	1.00	1.00	–	1.00	1.00	–
Investment	4.77	5.27	10 %	1.00	.98	2 %
Consumption	.44	.41	7 %	1.00	.92	8 %
Employment	.71	.73	3 %	.91	.91	0 %
Productivity	.47	.45	4 %	.77	.74	4 %

TABLE 22. THE PREDICTION SUCCESS OF THE TWO VERSIONS OF MINIMAL DYNAMICS COMPARED WITH THAT OF STOCHASTIC RBC MODELS.

Standard deviation proportions:

Variable	*Non – st*	*Stoch*	*D – D*	*H – R*	*K – P*	*empirical*
Investment	4.77	5.27	3.45	3.24	3.07	4.89
Consumption	.44	.41	.19	.29	.25	.73
Employment	.71	.73	.72	.77	.68	.94
Productivity	.47	.45	.35	.28	.40	.67
$\sum\Delta^2$	.1961	.3393	2.5185	3.1071	3.6933	

Correlations with output:

Variable	*Stoch*	*Non – st*	*K – P*	*H – R*	*D – D*	*empirical*
Investment	.98	1.00	.86	.99	.99	.92
Consumption	.92	1.00	.85	.87	.69	.85
Employment	.91	.91	.95	.98	.98	.76
Productivity	.74	.77	.86	.87	.91	.42
$\sum\Delta^2$	.1334	.1739	.2333	.2562	.3190	

Final Comments

There are three of them:

1. There can be no doubt about the testimony given by Figures 16 and 17 and, in a numerical form, by Table 22. They confirm the better quantitative success of the minimal dynamics constructed in Chapter 3 in the prediction of what are generally considered as important data on business cycles, viz. standard deviations and correlations with output of economic variables over a detrended cycle. In particular, these Figures and that Table add to the evidence given in Chapter 8 the remark that both the stochastic and nonstochastic versions of the minimal dynamics predict better than do the stochastic RBC models. Table 21 adds that the effect of shocks is rather small.

2. These results thus give complementary evidence for the effects of nonmaterial values on the business cycles, produced by the interaction between material and nonmaterial values.

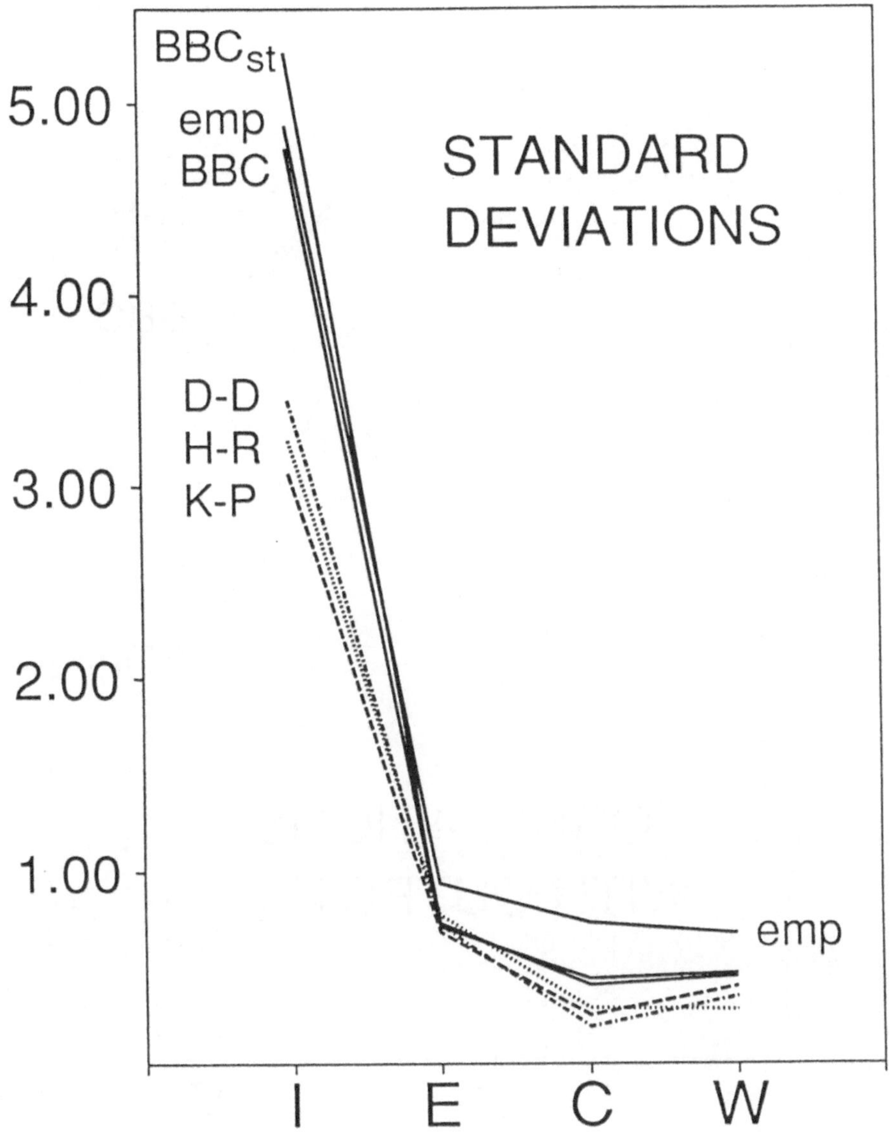

FIGURE 16: STANDARD DEVIATION PROPORTIONS. TWO DISTINCT PATTERNS APPEAR: THE PREDICTIONS OF MINIMAL DYNAMICS, BOTH STOCHASTIC (*BBC_{st}* ABOVE) AND NONSTOCHASTIC (*BBC* ABOVE) VERSIONS, FOLLOW THE PATTERN OF EMPIRICAL VALUES, WHILE THOSE OF THE RBC MODELS STRAY AWAY.

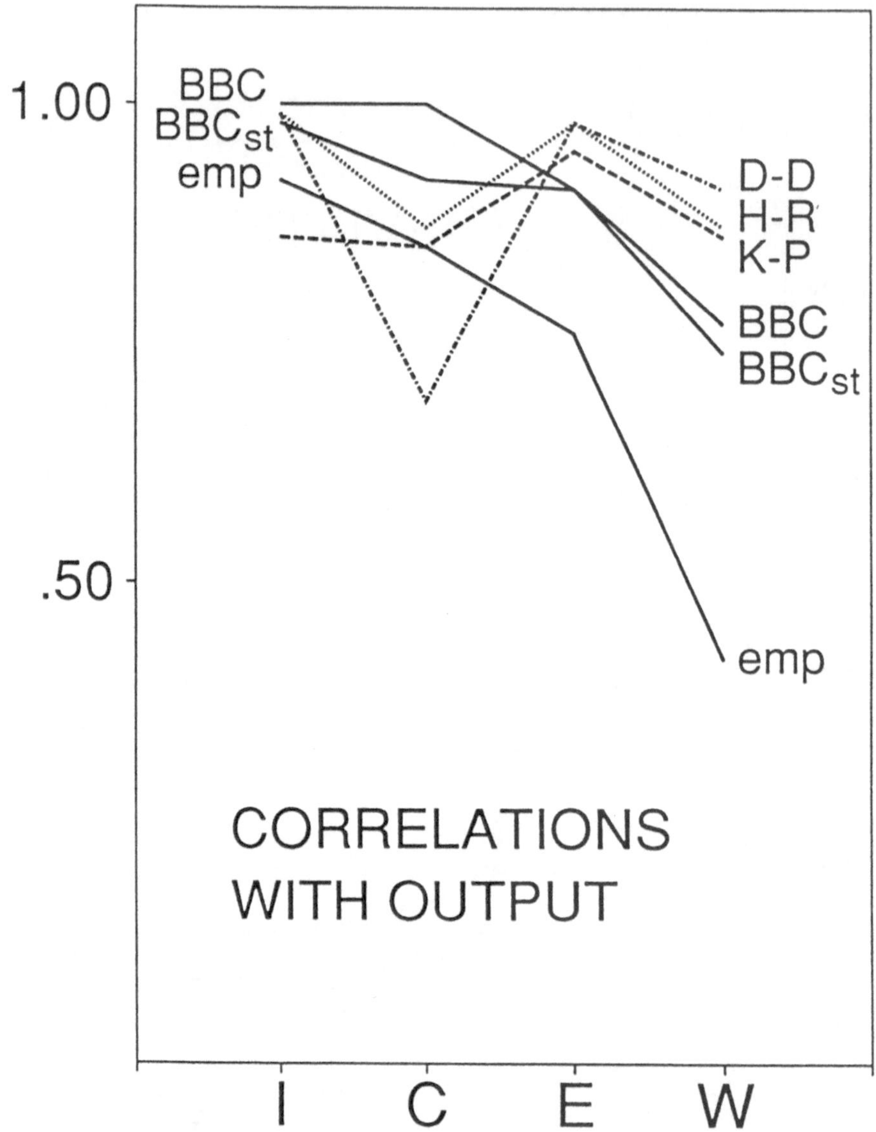

FIGURE 17: CORRELATIONS WITH OUTPUT.
AGAIN TWO PATTERNS APPEAR: THE PREDICTIONS OF MINIMAL DYNAMICS, BOTH STOCHASTIC (BBC_{st} ABOVE) AND NONSTOCHASTIC (BBC ABOVE), FOLLOW THE PATTERN OF EMPIRICAL VALUES, WHILE THOSE OF THE STOCHASTIC OPTIMIZATION MODELS TOGETHER STRAY OFF THIS PATTERN.

3. The decisive effect of the pursuit of cognitive innovations, also without any material rewards (e.g. in science), upon economic growth is obvious and widely recognized. But the most important nonmaterial values, like individual freedom and the pursuit of objective knowledge, can be represented only by an unbounded value function, of which the generalized value function constructed in Chapter 3 gives an example. This implies that in macroeconomic theory we should have two levels.

First of them deals with the always finite material values and bounded value functions. This level of macroeconomics is reducible to microeconomics and may be based on stochastic optimization (and possibly profits also from applications of game theory). It represents the level on which the economic game is played in the short run. This level, in other words, stands for macroeconomics as understood in economics today.

The second level represents the long-term economic development including business cycles. It is dominated by the continuous interaction between material and nonmaterial values. The nonmaterial values express the long-term human pursuits of ever larger individual freedom and ever greater objective knowledge of the world we are living in. This accordingly is the suggested new level of economics on which nonmaterial values are the decisive ones. This level of long-term economic development has been the object of study in this book.

Appendix (Chapter 10)

TABLE 23.
THE RANDOM SERIES $u_i(n)$ AND $x_1(n)$, AND THE CORRESPONDING VALUES OF THE FUNCTIONS Q_Y/x_o AND R.[1]

n	$u_1(n)$	$u_2(n)$	$x_1(n)$	$Q_Y(n)/x_o$	$R(n)$
0	.1415926	.7320508	−1.9647102	.0000000	1.0000000
1	.8848542	.0601036	.1823869	−.0045158	1.0033915
2	.4987814	.0528412	.3844457	−.0090623	1.0063630
3	.1137038	.7506604	−2.0852436	−.0136377	1.0089105
4	.0441046	.9593468	−.6312724	−.0182400	1.0110295
5	.9811582	.1129356	.1270773	−.0228676	1.0127159
6	.0271494	.8005852	−2.5511509	−.0275184	1.0139661
7	.6063598	.7855084	−.9754896	−.0321907	1.0147761
8	.2160566	.0061628	.0677678	−.0368825	1.0151426
9	.4899422	.9536076	−.3432899	−.0415918	1.0150621
10	.3116774	.2936092	1.1266953	−.0463168	1.0145315
11	.8017358	.0741164	.2985173	−.0510554	1.0135476
12	.1502486	.4948988	.0623949	−.0558056	1.0121079
13	.6288062	.8829196	−.6464034	−.0605654	1.0102093
14	.3315654	.8855132	−.9790354	−.0653328	1.0078497
15	.1062318	.7076844	−2.0432002	−.0701057	1.0050266
16	.6754806	.3359548	.7597361	−.0748820	1.0017381
17	.1273502	.4976716	.0296999	−.0796597	.9979820
18	.3700134	.7618972	−1.4061796	−.0844366	.9937568
19	.2942478	.5214124	−.2098076	−.0892107	.9890610
20	.2765526	.2877308	1.5584978	−.0939797	.9838931
21	.6671742	.2106636	.8723315	−.0987416	.9782521
22	.4942214	.7803612	−1.1657046	−.1034941	.9721372
23	.6681838	.3745004	.6369640	−.1082352	.9655475
24	.8142646	.7166268	−.6270039	−.1129625	.9584825
25	.1218782	.1706956	1.8022195	−.1176740	.9509419

[1] For the construction of the random series see Section 10.2; for the functional expressions of Q_Y and R see the equations (10.16) and (10 .20), respectively.

n	$u_1(n)$	$u_2(n)$	$x_1(n)$	$Q_Y(n)/x_o$	$R(n)$
26	.6353894	.1105052	.6093982	−.1223675	.9429256
27	.4184398	.0301484	.2485553	−.1270407	.9344338
28	.6454166	.5570428	−.3282650	−.1316913	.9254668
29	.5970622	.5825676	−.5035684	−.1363172	.9160251
30	.2687174	.6739292	−1.4394879	−.1409162	.9061096
31	.1834158	.6355564	−1.3857575	−.1454860	.8957209
32	.1428086	.4713788	.3528892	−.1500244	.8848606
33	.2703262	.4270796	.7154284	−.1545291	.8735298
34	.6934054	.3842332	.5689930	−.1589979	.8617303
35	.8095118	.8019244	−.6158205	−.1634286	.8494639
36	.6152406	.2100348	.9547266	−.1678190	.8367325
37	.0312702	.5810316	−1.2831537	−.1721668	.8235386
38	.9126534	.1870172	.3945056	−.1764696	.8098847
39	.3111278	.2844524	1.4924406	−.1807255	.7957732
40	.6275126	.1714108	.8500693	−.1849320	.7812074
41	.9214942	.3372236	.3451510	−.1890871	.7661905
42	.1136614	.8998812	−1.2270501	−.1931884	.7507256
43	.0306638	.2623404	2.6320198	−.1972337	.7348164
44	.7204246	.1619068	.6889139	−.2012211	.7184669
45	.3745982	.3244556	1.2507874	−.2051480	.7017345
46	.7476294	.8524252	−.6101264	−.2090125	.6844630
47	.9985198	.2187884	.0533864	−.2128123	.6668175
48	.5307766	.3559228	.8853319	−.2165453	.6487490
49	.2561822	.8275276	−1.4584093	−.2202095	.6302628
50	.2097574	.3262492	1.5683945	−.2238024	.6113638
51	.4930958	.4209964	.5663474	−.2273223	.5920575
52	.3113686	.4558588	.4182644	−.2307670	.5723494
53	.7038462	.5072396	−.0381095	−.2341342	.5522454
54	.1192454	.7949532	−1.9805970	−.2374222	.5317515
55	.8007918	.0001644	.0006885	−.2406288	.5108740
56	.8510006	.0521148	.1827011	−.2437519	.4896195
57	.7671902	.5203916	−.0930239	−.2467897	.4679945
58	.1992934	.9641372	−.4013022	−.2497401	.4460059
59	.1760078	.6314924	−1.4452538	−.2526012	.4236610
60	.7944726	.1830908	.6192787	−.2553711	.4009669

n	$u_1(n)$	$u_2(n)$	$x_1(n)$	$Q_Y(n)/x_o$	$R(n)$
61	.8478142	.0397836	.1421450	−.2580479	.3779312
62	.7571014	.6114012	−.4805603	−.2606297	.3545617
63	.0011438	.8141804	−3.3853679	−.2631149	.3309236
64	.3625846	.0951868	.8020293	−.2655014	.3068532
65	.9393182	.1742156	.3144772	−.2677875	.2825306
66	.7638694	.2263452	.7258813	−.2699717	.2579071
67	.1465998	.7514284	−1.9595377	−.2720519	.2329914
68	.4721366	.2028028	1.1716657	−.2740268	.2077926
69	.6673022	.2884876	.8732286	−.2758945	.1823197
70	.5347974	.4505692	.3419233	−.2776535	.1565818
71	.5307758	.8304364	−.9848241	−.2793023	.1305886
72	.2559286	.2483388	1.6508835	−.2808393	.1043499
73	.1293662	.7233996	−1.9942463	−.2822630	.0778753
74	.0090854	.3176732	2.7932558	−.2835719	.0511748
75	.8800718	.7024044	−.4830387	−.2847648	.0242588
76	.9827606	.6621948	−.1588238	−.2858402	−.0028622
77	.5351102	.9157516	−.5647015	−.2867967	−.0301782
78	.6299334	.2932572	.9261041	−.2876331	−.0576782
79	.6888878	.9625324	−.2013715	−.2883481	−.0853514
80	.3774326	.1227708	.9731757	−.2889407	−.1131866
81	.6461342	.9183436	−.4587510	−.2894094	−.1411729
82	.8245414	.1149212	.4105582	−.2897535	−.1692990
83	.3796238	.4300204	.5924445	−.2899716	−.1975532
84	.3407446	.3164668	1.3412791	−.2900630	−.2259242
85	.0160382	.3199756	2.6015556	−.2900265	−.2544001
86	.0841094	.4322652	.9186701	−.2898614	−.2829691
87	.6626798	.0280684	.1591564	−.2895665	−.3116192
88	.0694966	.8976828	−1.3844398	−.2891413	−.3403383
89	.0304222	.5654476	−1.0564577	−.2885851	−.3691142
90	.6438374	.2468892	.9382331	−.2878970	−.3979346
91	.0964558	.2683764	2.1545009	−.2870764	−.4267868
92	.5764886	.6488188	−.8445187	−.2861227	−.4556587
93	.7468862	.6755596	−.6819377	−.2850354	−.4845374
94	.7629254	.1523932	.6015928	−.2838140	−.5134100
95	.8473518	.3086444	.5369349	−.2824580	−.5422640

n	$u_1(n)$	$u_2(n)$	$x_1(n)$	$Q_Y(n)/x_o$	$R(n)$
96	.6105206	.8402748	−.8378531	−.2809672	−.5710863
97	.5350302	.3671116	.8290528	−.2793412	−.5998639
98	.6045734	.3743772	.7121589	−.2775796	−.6285838
99	.6497678	.6775724	−.8340856	−.2756825	−.6572328
100	.9763926	.7904508	−.2115664	−.2736495	−.6857976
101	.5164542	.5729036	−.5083620	−.2714806	−.7142653
102	.7159814	.6104412	−.5227975	−.2691758	−.7426222
103	.9661038	.5098604	−.0162599	−.2667352	−.7708552
104	.2549046	.6257468	−1.1746033	−.2641587	−.7989508
105	.8047582	.3617356	.5032416	−.2614466	−.8268954
106	.1083494	.6701852	−1.8486685	−.2585992	−.8546758
107	.3467598	.4487084	.4609675	−.2556167	−.8822784
108	.9228566	.2405628	.3999986	−.2524993	−.9096898
109	.5455422	.2584076	1.0993503	−.2492476	−.9368962
110	.9368774	.9152092	−.1834148	−.2458620	−.9638843
111	.9901358	.1213164	.0972340	−.2423432	−.9906405
112	.8730486	.4572988	.1381351	−.2386915	−1.0171513
113	.7564062	.9637196	−.1688646	−.2349077	−1.0434032
114	.7807654	.4991132	.0039199	−.2309925	−1.0693825
115	.5026318	.2188844	1.1505976	−.2269467	−1.0950761
116	.3342806	.3863548	.9695100	−.2227711	−1.1204703
117	.9669502	.4744716	.0414073	−.2184668	−1.1455516
118	.5232134	.4074972	.6249222	−.2140346	−1.1703070
119	.8586478	.1766124	.4944212	−.2094755	−1.1947227
120	.1913526	.9861308	−.1582765	−.2047908	−1.2187860
121	.6587742	.6034636	−.5529849	−.1999814	−1.2424830
122	.8314214	.2979612	.5802661	−.1950488	−1.2658008
123	.5605838	.4537004	.3085919	−.1899941	−1.2810977
124	.7050646	.8230268	−.7495486	−.1848187	−1.3112481
125	.5054782	.8994956	−.6895942	−.1795242	−1.3333509
126	.2365894	.1401052	1.3089680	−.1741117	−1.3550230
127	.9988398	.4133484	.0249569	−.1685831	−1.3762514
128	.6322166	.0314428	.1879610	−.1629399	−1.3970237
129	.4126622	.9673676	−.2708938	−.1571837	−1.4173272
130	.8139174	.6555292	−.5319428	−.1513164	−1.4371500

n	$u_1(n)$	$u_2(n)$	$x_1(n)$	$Q_Y(n)/x_o$	$R(n)$
131	.0118158	.8027564	−2.8171779	−.1453396	−1.4564793
132	.7456086	.4737788	.1256681	−.1392552	−1.4753034
133	.3579262	.1878796	1.3256648	−.1330653	−1.4936100
134	.4626054	.5578332	−.4413322	−.1267716	−1.5113877
135	.6459118	.8331244	−.8103280	−.1203763	−1.5286238
136	.7540406	.1004348	.4433272	−.1138815	−1.5453078
137	.0308702	.8378316	−2.2458848	−.1072893	−1.5614277
138	.7858534	.5926172	−.3815810	−.1006020	−1.5769728
139	.1155278	.8596524	−1.6037189	−.0938218	−1.5919316
140	.6223126	.5098108	−.0600006	−.0869510	−1.6062935
141	.2730942	.6100236	−1.0271840	−.0799921	−1.6200477
142	.5708614	.3774812	.7369748	−.0729475	−1.6331838
143	.9630638	.6615404	−.2330583	−.0658196	−1.6456914
144	.2912246	.7083068	−1.5171801	−.0586111	−1.6575605
145	.3181982	.5332556	−.3139155	−.0513246	−1.6687814
146	.8688294	.0420252	.1384049	−.0439626	−1.6793442
147	.4189198	.3219884	1.1864906	−.0365280	−1.6892394
148	.7975766	.0703228	.2876013	−.0290235	−1.6984581
149	.8317822	.2923276	.5855960	−.0214519	−1.7069912
150	.6749574	.6678492	−.7711650	−.0138161	−1.7148298
151	.9614958	.7081964	−.2706210	−.0061190	−1.7219655
152	.7941686	.4982588	.0074272	.0016363	−1.7283902
153	.7514462	.9480396	−.2424499	.0094470	−1.7340958
154	.2084454	.5285532	−.3160103	.0173100	−1.7390746
155	.0771918	.5513644	−.7178531	.0252222	−1.7433191
156	.4698006	.7825148	−1.2036219	.0331803	−1.7468220
157	.9267902	.0571916	.1371281	.0411813	−1.7495768
158	.7924934	.1297372	.4963946	.0492218	−1.7515765
159	.2204078	.1266924	1.2427530	.0572987	−1.7528150
160	.8692726	.1614908	.4495705	.0654085	−1.7532860
161	.5594142	.1925836	1.0084550	.0735479	−1.7529839
162	.3343014	.0490012	.4486076	.0817136	−1.7519021
163	.9735438	.5333804	−.0482130	.0899021	−1.7500387
164	.6133846	.0815868	.4849238	.0981100	−1.7473857
165	.4429182	.8630156	−.9677559	.1063337	−1.7439396

n	$u_1(n)$	$u_2(n)$	$x_1(n)$	$Q_Y(n)/x_o$	$R(n)$
166	.4050694	.5759452	−.6174449	.1145698	−1.7396961
167	.4069998	.5746284	−.6059431	.1228146	−1.7346516
168	.0189366	.1572028	2.3512551	.1310646	−1.7288022
169	.0029022	.8332876	−2.9607997	.1393161	−1.7221451
170	.9199974	.1521692	.3336217	.1475656	−1.7146768
171	.6391758	.2376364	.9432710	.1558094	−1.7063954
172	.6187286	.3307388	.8564802	.1640436	−1.6972982
173	.1369662	.8441996	−1.6548154	.1722648	−1.6873836
174	.4182854	.6112732	−.8496997	.1804690	−1.6766499
175	.5964718	.7736044	−1.0054256	.1886526	−1.6650962
176	.0815606	.2325948	2.2255571	.1968118	−1.6527214
177	.8547102	.7325516	−.5569801	.2049428	−1.6395255
178	.9431334	.2188572	.3356612	.2130418	−1.6255077
179	.9732878	.3777324	.1616970	.2211049	−1.6106688
180	.5322326	.7411708	−1.1213703	.2291284	−1.5950095
181	.7177342	.9511436	−.2461035	.2371084	−1.5785304
182	.5217414	.5125212	−.0896486	.2450409	−1.5612331
183	.3920238	.4692204	.2630179	.2529223	−1.5431194
184	.2715446	.7428668	−1.6130788	.2607486	−1.5241901
185	.0796382	.4887756	.1585186	.2685160	−1.5044513
186	.2453094	.9418652	−.5988315	.2762207	−1.4839026
187	.7630798	.5712684	−.3184028	.2838586	−1.4625473
188	.8962966	.0920828	.2558835	.2914260	−1.4403913
189	.1260222	.1902476	1.8935711	.2989190	−1.4174366
190	.9490374	.3084892	.3018445	.3063337	−1.3936882
191	.8448558	.7910764	−.5614384	.3136664	−1.3691509
192	.8192886	.7712188	−.6257746	.3209132	−1.3438299
193	.7144862	.4763596	.1213514	.3280701	−1.3177307
194	.4921254	.0059932	.0448311	.3351335	−1.2908594
195	.0037518	.8998444	−1.9672033	.3420994	−1.2632223
196	.1893206	.2506748	1.8244359	.3489642	−1.2348259
197	.0146302	.4639116	.6534775	.3557241	−1.2056774
198	.6377734	.0599772	.3490188	.3623752	−1.1757840
199	.1741678	.0127724	.1498783	.3689139	−1.1451542
200	.8765046	.5585468	−.1846452	.3753365	−1.1137957

n	$u_1(n)$	$u_2(n)$	$x_1(n)$	$Q_Y(n)/x_o$	$R(n)$
201	.2111926	.0488508	.5328300	.3816392	−1.0817174
202	.9480542	.4857036	.0293007	.3878185	−1.0489282
203	.5331814	.9680412	−.2236924	.3938706	−1.0154374
204	.0185038	.8690604	−2.0705787	.3997920	−.9812549
205	.8657046	.4921468	.0264889	.4055792	−.9463910
206	.4283582	.0105356	.0861354	.4112285	−.9108557
207	.7895494	.3397852	.5809287	.4167365	−.8746604
208	.2871598	.7119084	−1.5346653	.4220997	−.8378161
209	.0296566	.6749628	−2.3631776	.4273147	−.8003344
210	.4011422	.9632076	−.3096823	.4323782	−.7622276
211	.1620774	.3368092	1.6309022	.4372868	−.7235076
212	.3785358	.7685164	−1.3844528	.4420371	−.6841876
213	.9958486	.6196988	−.0623146	.4466259	−.6442802
214	.6840062	.4445196	.2976963	.4510503	−.6037993
215	.8299654	.9127132	−.3183015	.4553067	−.5627583
216	.0990318	.3300844	1.8839432	.4593922	−.5211716
217	.3930806	.6367548	−1.0349719	.4633038	−.4790536
218	.6065502	.8512716	−.8042689	.4670384	−.4364189
219	.2764134	.8530972	−1.2787976	.4705932	−.3932831
220	.6230478	.4318124	.4041316	.4739653	−.3496610
221	.5061526	.8845308	−.7743125	.4771518	−.3055693
222	.4503742	.3962636	.7761981	.4801499	−.2610233
223	.7686214	.6155612	−.4816808	.4829571	−.2160399
224	.6529838	.1329004	.6844315	.4855707	−.1706358
225	.9958646	.1294268	.0661390	.4879882	−.1248281
226	.6890782	.0282956	.1526259	.4902069	−.0786340
227	.4377894	.9697052	−.2431820	.4922246	−.0320714
228	.7792398	.3965484	.4274518	.4940388	.0148416
229	.0190166	.7058428	−2.7074605	.4956474	.0620873
230	.0282622	.7521676	−2.6704187	.4970480	.1096469
231	.9591174	.4371292	.1111923	.4982385	.1575019
232	.0402158	.5699564	−1.0787863	.4992171	.2056326
233	.7484086	.6761788	−.6808847	.4999816	.2531074
234	.2455262	.3486796	1.3639723	.5005301	.3026449
235	.8318054	.5314332	−.1190830	.5008610	.3514867

n	$u_1(n)$	$u_2(n)$	$x_1(n)$	$Q_Y(n)/x_o$	$R(n)$
236	.6823118	.4643244	.1943601	.5009725	.4005260
237	.2928406	.1908348	1.4601925	.5008528	.4497422
238	.8304702	.4946316	.0205559	.5005305	.4991147
239	.2590534	.7982172	−1.5687536	.4999742	.5486232
240	.1199278	.0348524	.4474116	.4991923	.5982464
241	.0171126	.0482108	.8508720	.4981838	.6479629
242	.4246942	.2828236	1.2809949	.4969472	.6977519
243	.6280614	.6550812	−.7979846	.4954817	.7475916
244	.0954638	.6607404	−1.8354558	.4937861	.7974606
245	.2620246	.4547068	.4595062	.4918596	.8473365
246	.0617982	.1420556	1.8373850	.4897013	.8971974
247	.5900294	.0316252	.2027737	.4873105	.9470213
248	.0393198	.0251884	.4009469	.4846866	.9967858
249	.4643766	.9847228	−.1187095	.4818291	1.0464685
250	.2073822	.1571276	1.4802762	.4787375	1.0960468
251	.7401574	.8094492	−.7222561	.4754116	1.1454976
252	.6298958	.5953964	−.5423985	.4718510	1.1947986
253	.6769686	.7406588	−.8818026	.4680557	1.2439266
254	.5990462	.7888396	−.9823453	.4640257	1.2928587
255	.8976454	.0621532	.1769028	.4597611	1.3415719
256	.5535918	.7025644	−1.0395538	.4552621	1.3900428
257	.4886006	.7129148	−1.1644928	.4505288	1.4382483
258	.8863902	.9939916	−.0185361	.4455619	1.4861650
259	.9856934	.0953372	.0957188	.4403617	1.5337699
260	.4648078	.2218924	1.2185920	.4349288	1.5810393
261	.3440726	.3398908	1.2338928	.4292642	1.6279502
262	.0710142	.7453836	−2.2989782	.4233684	1.6744789
263	.5115014	.2866012	1.1274481	.4172426	1.7206023
264	.1459438	.8525804	−1.5683091	.4108877	1.7662969
265	.2641846	.2679868	1.6212204	.4043048	1.8115393
266	.7465182	.9518156	−.2279749	.3974953	1.8563063
267	.6462694	.7255452	−.9233776	.3904606	1.9005749
268	.8673998	.9978284	−.0072776	.3832021	1.9443215
269	.9657366	.3116028	.2445264	.3757214	1.9875229
270	.1385022	.7780876	−1.9575166	.3680203	2.0301561

n	$u_1(n)$	$u_2(n)$	$x_1(n)$	$Q_Y(n)/x_o$	$R(n)$
271	.9051974	.6537692	−.3671941	.3601005	2.0721985
272	.9475758	.2448364	.3279989	.3519640	2.1136268
273	.3815286	.6131388	−.9058025	.3436129	2.1544181
274	.9445662	.3649996	.2533327	.3350492	2.1945502
275	.4274854	.7048732	−1.2516581	.3262752	2.2340004
276	.5128718	.4448044	.3927883	.3172933	2.2727462
277	.5803606	.0029948	.0196280	.3081060	2.3107652
278	,9743102	.9493516	−.0713846	.2987158	2.3480356
279	.8563334	.9444572	−.1904454	.2891255	2.3845352
280	.4576878	.3929324	.7790584	.2793379	2.4202427
281	.0870326	.5595708	−.8079153	.2693557	2.4551361
282	.5893342	.3839436	.6851714	.2591822	2.4891944
283	.8189414	.7101212	−.6123111	.2488203	2.5223960
284	.6044238	.1084204	.6319328	.2382733	2.5547207
285	.6023446	.3692668	.7371672	.2275447	2.5861473
286	.9432382	.0575756	.1209931	.2166377	2.6166555
287	.0065094	.2514652	3.1730366	.2055559	2.6462253
288	.0634798	.7144684	−2.2899286	.1943031	2.6748370
289	.1230966	.4864828	.1736317	.1828828	2.7024706
290	.0216222	.2150476	2.7026187	.1712991	2.7291067
291	.8542374	.1700892	.4920493	.1595558	2.7547269
292	.7932558	.9182764	−.3343220	.1476569	2.7793120
293	.4620886	.0936188	.6894867	.1356067	2.8028438
294	.4820862	.6771596	−1.0836782	.1234093	2.8253043
295	.8213254	.6595932	−.5288978	.1110691	2.8466756
296	.3601518	.0910444	.7736766	.0985906	2.8669406
297	.1681206	.8610748	−1.4468938	.0859782	2.8860823
298	.2942302	.9607116	−.3822277	.0732366	2.9040840
299	.2709734	.5455772	−.4564757	.0603705	2.9209295
300	.8985678	.9479724	−.1485123	.0473847	2.9366031
301	.8459926	.5072508	−.0263394	.0342841	2.9510896
302	.1796542	.7985036	−1.7675705	.0210736	2.9643733
303	.9503814	.1256412	.2264993	.0077584	2.9764404
304	.2709038	.8282604	−1.4246798	−.0082690	2.9862545

TABLE 24.
CALCULATION OF THE SUMS OVER n OF THE PRODUCTS $x_1(n)Q_Y(n)/x_o$ AND $x_1(n)R(n)$.

N	$\sum_{n=0}^{N} x_1(n)Q_Y(n)/x_o$	N	$\sum_{n=0}^{N} x_1(n)R(n)$
10	.0939377	10	−6.7214615
20	.2532940	20	−9.3297653
30	.2830836	30	−9.1443151
40	−.1044807	40	−7.5678072
50	−.9524426	50	−4.6857731
60	−.4131410	60	−5.6820003
70	−.0507954	70	−6.3972300
80	−.6131496	80	−6.5065263
90	−1.7920221	90	−7.4149483
100	−2.2029405	100	−7.9924522
110	−1.7108395	110	−6.7500003
120	−2.4196728	120	−10.3377010
130	−2.3582360	130	−9.8003370
140	−1.6046544	140	.1159521
150	−1.5155529	150	1.5962746
160	−1.4489933	160	2.2974635
170	−1.5980787	170	3.2557601
180	−1.7645042	180	4.3010455
190	−1.7715406	190	4.5247972
200	−1.7850229	200	4.9845704
210	−3.9894607	210	9.4378826
220	−4.3266594	220	9.4712318
230	−6.6992265	230	9.0280998
240	−6.6217173	240	9.1742093
250	−4.7925534	250	12.686858
260	−6.6131011	260	8.0079055
270	−7.6577445	270	2.5933738
280	−8.0373681	290	.5358483
290	−7.1832264	290	12.6671250
300	−7.3637752	300	5.6213579
304	−7.3883893	304	−3.2764056

Referred Literature

ARROW,K.J. AND M.KURZ, (1970), Public Investment, the Rate of Return and Optimal Fiscal Policy (John Hopkins Press, Baltimore, Maryland).

AULIN,A. (1996), Causal and Stochastic Elements in Business Cycles, Lecture Notes in Economics and Mathematical Systems 431, (Springer-Verlag, Berlin).

AULIN,A. (1997), The Origins of Economic Growth, (Springer-Verlag, Berlin).

AULIN-AHMAVAARA,P. (1997), Measuring the productivity of nations, in A.Simonovits and E.Steenge (eds.), Prices, Growth and Cycles (Macmillan, London).

BARRO,R. AND X.SALA-I-MARTIN (1995), Economic Growth (McGraw-Hill, New York).

BROCK,W.A. AND L.J.MIRMAN (1972), Optimal economic growth and uncertainty: the discounted case, Journal of Economic Theory 4, 497–513.

COOLEY,T.F. (ed.) (1995), Frontiers of Business Cycle Research, (Princeton University Press, Princeton).

COOLEY,T.F. AND E.C.PRESCOTT, Economic growth and business cycles, in Cooley,T.F. (ed.), Frontiers of Business Cycle Research, 1-38 (Princeton University Press, Princeton).

CORREIA,I., J.NEVES AND S.REBELO (1992), Business cycles from 1850 to 1950, European Economic Review 36, 459-467.

COURANT,R. AND D.HILBERT (1953), Methods of Mathematical Physics I-II (Interscience Publishers, New York).(First German ed. in 1924.)

DANTHINE,J.P. AND J.DONALDSON (1990), Efficiency wages and the business cycle puzzle, European Economic Review 34, 1275-1301.

DANTHINE,J.P. AND J.DONALDSON (1991), Risk sharing, the minimum wage and the business cycles, in: W.Barnett, B.Cornet, G.D'Aspremont, J.Gabszewicz and A.Mas-Collell (eds.), Equilibrium Theory and Applications (Cambridge University Press, New York).

DANTHINE,J.P. AND J.DONALDSON (1992), Risk sharing in the business cycle, European Economic Review 36, 468-475.

DANTHINE,J.P. AND J.DONALDSON (1993), Methodological and empirical issues in real business cycle theory, European Economic Review 37, 1-35.

DANTHINE,J.P. AND M.GIRARDIN (1989), Business cycles in Switzerland: A comparative study, European Economic Review 33, 31-50.

DAY,R.H. (1992), Review article, in Structural Change and Economic Dynamics 3, 177-182.

DENISON,E.F. (1961), The Sources of Economic Growth in the United States (Committee for Economic Development, New York).

DENISON,E.F. (1983), The interruption of productivity growth in the United States, The Economic Journal, March.

EVANS,G., S.HONKAPOHJA AND P.M.ROMER (1996), Growth cycles, manuscript dated September 26, homepage of Paul Romer.

GANDOLFO,G. (1996), Economic Dynamics (Springer-Verlag, Berlin).

HANSEN,G. (1985), Indivisible labor and the business cycle, Journal of Monetary Economics 16, 309-328.

JORGENSON,D.W., F.M.GOLLOP AND B.FRAUMENI (1987), Productivity and U.S. Economic Growth (North-Holland, Amsterdam).

KURZ M. (1968), The general instability of a class of competitive growth processes, Review of Economic Studies 35, 155.

KYDLAND,F.E. (1995),Business cycles and aggregate labor market fluctuations, in Thomas F. Cooley (ed.), Frontiers of Business Cycle Research, 126-156 (Princeton University Press, Princeton).

KYDLAND,F. AND E.C.PRESCOTT (1982), Time to build and aggregate fluctuations, Econometrica 50, 1345-1370.

KYDLAND,F. AND E.C.PRESCOTT (1990), Business cycles: real facts and monetary myth, Federal Reserve Bank of Minneapolis Quarterly Review 14, 3-18.

LUCAS,R.E.JR. (1987), Models of Business Cycles (Basil Blackwell, London).

LUCAS,R.E.JR. (1988), On the mechanics of economic development, Journal of Monetary Economics 22, 3-42.

MADDISON,A. (1982), Phases of Capitalist Development (Oxford University Press, Oxford).

MADDISON,A. (1991), Dynamic Forces in Capitalist Development (Oxford University Press, Oxford).

MANKIW,G., D.ROMER AND D.WEIL (1990), A contribution to the empirics of economic growth (NBER Working Paper Series No. 3541, Cambridge, MA).

MIROWSKI,P. (1990), From Mandelbrot to chaos in economic theory, Southern Economic Journal 57, 289.

PLOSSER,C.I. (1992), The search for growth, Federal Reserved Bank of Kansas City Economic Review Symposium, 57-86.

PRESCOTT,E. (1986), Theory ahead of business cycle measurement, Federal Reserve Bank of Minneapolis Quartely Review 10, 9-22.

REBELO,S. (1991), Long-run policy analysis and long-run growth, Journal of Political Economy 99, 500-521.

ROGERSON,W. (1985), Indivisible labor and the business cycle (unpublished working paper, University of Rochester).

ROMER,P.M. (1986), Increasing returns and long-run growth, Journal of Political Economy 94, 1002-10037.

ROMER,P.M. (1987), Crazy explanations for the productivity slowdown, NBER Macroeconomics Annual 1987, 163-205.

ROMER,P.M. (1988), The origins of endogeneous economic growth, Journal of Economic Perspectives 8, 1988, 3-42.

SAMUELSON, P.A. AND W.D. NORDHAUS (1985), Economics (12th ed.), (McGraw-Hill, New York).

SLUTSKY, E. (1937), The summation of random causes as the cause of cyclic processes, Econometrica, 105-146.

SOLOW, R.M. (1956), The production function and the theory of capital, Review of Economic Studies 23, 101.

SOLOW, R.M. (1957) Technical change and the aggregate production function, The Review of Economics and Statistics, 312-320.

STOKEY, N. AND R.E.LUCAS WITH E.C.PRESCOTT (1989), Recursive Methods in Economic Dynamics (Harvard University Press, MA).

SUMMERS,L. AND A.HESTON (1991), The Penn World Table (Mark 5): An expanded set of international comparisons, in Quarterly Journal of Economics 106, 327-368.

VAN BATH,B.H.SLICHER (1963), Yield ratios 810-1820, Afdeling Agrarische Geschiedenis, Bijdragen No.10, Wageningen.

VAN DUIJN, J. (1981), Fluctuations in innovations over time, Futures 13, 264.

Index

Algorithm
— solving fundamental equations, 51-52
— defining ordinary growth path, 72-73
— defining anomalous growth path, 77-79
Anomalies in science
— significance in verification of general theories,150
— significance in verification of cycle theories,150-151
Anomalous business cycles
— crash effect, 68
— observations,149,158,160,162,163-165, 167-171
— theoretical predictions,158,160,162, 163-171
— theory predicts only part of observations in the U.K. economy ,165-166
— verification of theory in the U.S. case assessed good,162-163
— visual image,166
Anomalous growth path
— defined, 77-79
— verified, 80-84
Autocorrelations, 145-148

Barro,R., 2,90-91,93-94,96,99,114
Basic macroeconomic facts
— No.1-6 (Kaldor facts), 2,68,74-75
— No.7 (Plosser fact), 2-3,69,92,94,97, 99,101
— No. 8-9 (irregular variations), 3,44,71, 181-184
— No. 10-12 (Kydland facts), 3,69,105-106,117-118,120,123
— No. 13-15 (ordinary cycle correlations and standard deviations), 4,70,130-133, 136-139,144-145
— No. 16 (anomalous growth), 4-5,68,81-83
— No. 17 (anomalous cycles), 5,149,158, 160,162-165,167-171
Beauty as a nonmaterial value of the demand side,38
Bernoulli,J., 17
Brock,W., 62

Calibration
— anomalous cycles,153,156
— ordinary cycles in Canada 1947-73, 96
— ordinary cycles in France 1950-73, 99
— ordinary cycles in Japan 1950-73, 93-94
— ordinary cycles in USA 1947-73, 91, 114-115
— ordinary cycles in USA, first half of the century, 134
— note on calibration in Chapters 8 and 9, 129
— relative independence of predictions from calibration,146
— stochastic ordinary cycles,180
—'traditional calibration',127,129,134, 142
Canonical equations of motion, 22,24
Canonical momentum, 18,22
Canonical vs Hamiltonian system, 19
Causal explanation in economics
— dicussed by Richard Day, 83
— realized by cycle functions, 82,84
Clockwise revolution
— defined, 47
— defining anomalous business cycles, 47
Complementary evidence discussed, 70-71

Condition on constant m,54
Continuous invariance group,43-45
Convergent vs divergent cycles, 148
Cooley, T.,3n, 62n,70,105n,124,138-140, 144-145
Cooley-Prescott model,138-140,144-145
Copernicus,N., 8
Correia,I., 5,149n,160,162
Counterclockwise revolution
— defined, 47
— defining ordinary business cycles, 47
Cycle center
— fixed, 46
— moving, 77-79
Cycle equations
— continuous invariance group, 43-44
— first normal form, 42
— invariance in time-inversion, 45-46
— second normal form, 45
— linear approximations, 49-50
Cycle functions
— convergent vs divergent, 148
— defined, 57
— of anomalous cycles, 151-152
— of hours in intellectual work,142
— of total working hours,141-142
— of various other economic variables, 58-61
— significance discussed, 61-63
Cycles
— convergent vs divergent, 148
— diminishing relative significance,147

Danthine,J., 70,130,180-181
Danthine-Donaldson model,130,180-181
Day,R., 6,62,173
Denison,E.,121,127,129
Decisive evidence discussed, 69-70
Dogmas of macroeconomics
— reduction to microeconomics, 5-6
— rejected in minimal dynamics, 67
— stochastic formulation, 6
— rejected in minimal dynamics,67,71
Donaldson,J., 70,130

Economic relevance of nonmaterial values, 6,11,38-39,
— decisive evidence discussed,124
Economics discussed
— an influential profession, 1
— macrotheory far from empirics, 1-2
Einstein,A., 1-3,9
Euler equation
— introduced, 16
— generalized, 18
— in Arrow-Kurz generalization, 28-29
— in minimal dynamics, 40
— in Lucasian mechanics, 31,33
— in Solow model, 24
— on balanced-growth path, 25,28,73
Evans,G.,2n,11

Fig.1: 36
Fig.2: 48
Fig.3: 83
Fig.4: 109
Fig.5: 110
Fig.6: 117
Fig.7: 120
Fig.8: 123
Fig.9: 131
Fig.10: 132
Fig.11: 136
Fig.12: 164
Fig.13: 169
Fig.14: 170
Fig.15: 171
Fig.16: 183
Fig.17: 184

Fraumeni,B., 90
Friedman,M., 67

Galileo, G., 8
Geodetic distance, 17
Generalized human capital
— defined, 9
— mathematical representation, 37
— immediately profitable part, 38
Generalized value function, 39
Golden Age period, 90,93,96,98,103
Gollop,F., 90
Growth paths
— appearance of two growth paths, 68-69
— balanced-growth path, 72-73
— logistic growth path, 77-80,
— verified, 80-83

Hamilton,W., 8,15,17
Hamilton-Jacobi equation, 19
Hamiltonian equations, 15
Hansen,G.,70,130,180-181
Hansen-Rogerson model,130,180-181
Heston,A., 90,129
Hidden variables, 37
Honkapohja,S.,2n,11
Hours
— in formal education 35-36,141
— of intellectual work 35-37,141-142
— of paid intellectual work 35-36,141
— of physical work 35-37,41,141
Human capital
— defined in economics, 7
— generalized, 9,37
— growth equation in Lucasian mechanics, 30
— growth equation in minimal dynamics, 37,51
— parameter indicating the immediately profitable part, 37
— positivity of its growth rate, 53-54

Individual freedom, 6,40
Inexhaustibility of nonmaterial values, 38
Intellectual beauty, 38
Interaction between material and nonmaterial values defined, 40
Invariance group,43-44
Irregular period 1974-82, 89,103-104,114

Jorgenson,D., 90

Kaldor,N., 2-4,68,74
Kaldor facts, 2-3,68
— derived from minimal dynamics, 74-75
Kepler,J., 8
Kurz.M.,22,24
Kurz momentum, 22-23
Kydland,F., 3-4,69-70,105-106,116,124, 130-131,133-137,144-145,180-181
Kydland facts, 3,69,105-106,116,124
Kydland-Prescott model,4,130-131, 133-137,180-181

Lagrangian function, 18
Legendre condition,20-21,23-24,28,31, 34,55
Legendre function, 18-19,22-24,
Legendre transformation, 18,21
Leisure, economic relevance discussed,38-39
Levels of macroeconomic theory
— long-term level,147,185
— short-term level, 147,185
Logistic function $b(t)$, 77
Logistic growth path, 77-78
Lucas,R.,25,28,30,32,39,51,56,62,73,127-129,133,150,153

Macroeconomics
— defined in two levels, 146-147
Maddison,A., 6-11,90,97-98

Malthus,T., 10
Mankiw,N., 127n,129
Market clearing 30,32
Markov process,172
Maximums and minimums in detrended cycles, formulae
— employment, 111
— output,consumption and investment,109
— productivity of labour, 112
— total hours, 113
Maximums and minimums in detrended cycles, numerical
— Golden Age period,115
— period 1950-81, 118-119
— first half of century, 121-122
Maximums and minimums in detrended cycles, predictions compared with empirics
— Golden Age period, 118
— period 1950-81, 120
— first half of century, 123
m-condition,54
Methods (a) and (b), 3,6,45,67
Mirman,L., 62
Mirowski,P., 19n
Modified Hamiltonian equations, 22-23,29

Natural boundary conditions, 18,23-24, 31,55,75
Neves,J., 5,149n,160,162
Newton,I., 1,8
Nonmaterial values
— as pursuits in leisure time, 67
— beauty in arts and fine arts, 38
— individual freedom, 6,39
— intellectual beauty in science, 38
— interaction with material values, 40
— nonreducibility to microeconomics, 67
— objective truth in science, 11,38
— 'purely' nonmaterial values, 38
— representation in generalized value function,39
— their economic relevance, 6,11,38-39
— their unbounded inexhaustible value, 38
— wanderings of thought in leisure, 38
Nontrivial macroeconomic facts, 3
Nordhaus,W., 5n,90,129
Note
— No.1: 3
— No.2: 3
— No.3: 3
— No.4: 4
— on errors in an earlier book by the author, 70
— on improvements to the predictions of Table 8:144-145
— on the relevance of the choice of normal form only in Chapters 5,7 and 9: 48

Objective truth, 9,38
OPEC oil shocks, 89
Ordinary business cycles
— defined by counterclockwise revolution, 47
— formulae for calculation of correlations and standard deviations over cycles, 126-127,142-143
— formulae for calculation of autocorrelations,134
— linear approximation of cycle functions, 126
— linear approximation of cycles, 48-50
— numerical calculations,128,135
— predictions compared with stochastic models, 130-132,136-137,139,143-145
Ordinary growth path

— defined, 72-73
— stochastic shocks added, 180-190

Phases 1 and 2
— in Lucasian mechanics, 30
— in minimal dynamics, 51
Physical capital discarded from comparisons, 132-133
Planck,M., 1
Plosser,C., 2-3,69,85,87,92,94,99-102
Plosser fact, 2-3,69,85,
— reproduced from theory, 87-88,92,94,97, 99-102
Prescott,E., 4,62n,70,130-131,133-140, 144-145,180-181
Purely nonmaterial values, 38

RBC-models, 4,61,70-71,181-182
Rebelo,S., 5,149n,160,162
Reduction to microeconomics discussed, 38
Risk aversion coefficient, 25,39,127
Rogerson,R., 70,130,180-181
Romer,D.,127n,129
Romer,P., 1n,2n,11

Sala-i-Martin,X., 2,90,93,99,114
Samuelson,P., 5n,90,129
Savings rate effects
— long-term, discussed 102-103
— short-term, theoretically produced, 92-95,97-98,100-101
Slicher van Bath,B., 8n
Slutsky,E., 61-62,172
Solow,R., 4,23,26-27,34,56,68,73,80-82, 84,121,150
Soviet economy, 5
Stochastic approach to business cycles
— brief history, 62
— criticized by Richard Day, 62-63
— failure to predict nontrivial facts, 63
— small effects on correlations and standard deviations,181
— weakness in prediction of basic facts generally,173
Stochastic cycle functions
— defined,176
— correlations and variances over detrended cycles,176-177
Stochastic optimization questioned in long-term development, 70-71,173
Stokey,N.,62n
Summers,R. 90,129

Table
— No.1: 81
— No.2: 103
— No.3: 106
— No.4: 118
— No.5: 120
— No.6: 123
— No.7: 128
— No.8: 130
— No.9: 133
— No.10: 135
— No.11: 137
— No.12: 139
— No.13: 158
— No.14: 160
— No.15: 162
— No.16: 162
— No.17: 163
— No.18: 165
— No.19: 167
— No.20: 168
— No.21: 181
— No.22: 182
— No.23: 186
— No.24: 195
Technological progress

— three Maddison categories, 11
— as total factor productivity, 37
Thesis, 11
Time inversion, 69-70
Time scale, 69
Total factor productivity
— defined, 37
— calculated, U.S.A., 91
— calculated, Japan, 94
— calculated, Canada, 96
— calculated, France, 99
Total human time, 35
Total hours
— cycle function defined,141-142
— total hours defined,35-36
Transversality conditions
— Courant-Hilbert form 20,55-56,75-76
— in minimal dynamics, 55-56,75-76
— in Lucasian mechanics, 32,34
— in Solow model, 26-27
— in terms of Kurz momentums, 23
Two levels of macroeconomic theory 147,185

Unbounded nonmaterial values, 38

Value function
— definition, 22
— generalized, 39

Wanderings of thought in leisure, 38
Weil,D.,127n,129

Vol. 367: M. Grauer, D. B. Pressmar (Eds.), Parallel Computing and Mathematical Optimization. Proceedings. V, 208 pages. 1991.

Vol. 368: M. Fedrizzi, J. Kacprzyk, M. Roubens (Eds.), Interactive Fuzzy Optimization. VII, 216 pages. 1991.

Vol. 369: R. Koblo, The Visible Hand. VIII, 131 pages. 1991.

Vol. 370: M. J. Beckmann, M. N. Gopalan, R. Subramanian (Eds.), Stochastic Processes and their Applications. Proceedings, 1990. XLI, 292 pages. 1991.

Vol. 371: A. Schmutzler, Flexibility and Adjustment to Information in Sequential Decision Problems. VIII, 198 pages. 1991.

Vol. 372: J. Esteban, The Social Viability of Money. X, 202 pages. 1991.

Vol. 373: A. Billot, Economic Theory of Fuzzy Equilibria. XIII, 164 pages. 1992.

Vol. 374: G. Pflug, U. Dieter (Eds.), Simulation and Optimization. Proceedings, 1990. X, 162 pages. 1992.

Vol. 375: S.-J. Chen, Ch.-L. Hwang, Fuzzy Multiple Attribute Decision Making. XII, 536 pages. 1992.

Vol. 376: K.-H. Jöckel, G. Rothe, W. Sendler (Eds.), Bootstrapping and Related Techniques. Proceedings, 1990. VIII, 247 pages. 1992.

Vol. 377: A. Villar, Operator Theorems with Applications to Distributive Problems and Equilibrium Models. XVI, 160 pages. 1992.

Vol. 378: W. Krabs, J. Zowe (Eds.), Modern Methods of Optimization. Proceedings, 1990. VIII, 348 pages. 1992.

Vol. 379: K. Marti (Ed.), Stochastic Optimization. Proceedings, 1990. VII, 182 pages. 1992.

Vol. 380: J. Odelstad, Invariance and Structural Dependence. XII, 245 pages. 1992.

Vol. 381: C. Giannini, Topics in Structural VAR Econometrics. XI, 131 pages. 1992.

Vol. 382: W. Oettli, D. Pallaschke (Eds.), Advances in Optimization. Proceedings, 1991. X, 527 pages. 1992.

Vol. 383: J. Vartiainen, Capital Accumulation in a Corporatist Economy. VII, 177 pages. 1992.

Vol. 384: A. Martina, Lectures on the Economic Theory of Taxation. XII, 313 pages. 1992.

Vol. 385: J. Gardeazabal, M. Regúlez, The Monetary Model of Exchange Rates and Cointegration. X, 194 pages. 1992.

Vol. 386: M. Desrochers, J.-M. Rousseau (Eds.), Computer-Aided Transit Scheduling. Proceedings, 1990. XIII, 432 pages. 1992.

Vol. 387: W. Gaertner, M. Klemisch-Ahlert, Social Choice and Bargaining Perspectives on Distributive Justice. VIII, 131 pages. 1992.

Vol. 388: D. Bartmann, M. J. Beckmann, Inventory Control. XV, 252 pages. 1992.

Vol. 389: B. Dutta, D. Mookherjee, T. Parthasarathy, T. Raghavan, D. Ray, S. Tijs (Eds.), Game Theory and Economic Applications. Proceedings, 1990. IX, 454 pages. 1992.

Vol. 390: G. Sorger, Minimum Impatience Theorem for Recursive Economic Models. X, 162 pages. 1992.

Vol. 391: C. Keser, Experimental Duopoly Markets with Demand Inertia. X, 150 pages. 1992.

Vol. 392: K. Frauendorfer, Stochastic Two-Stage Programming. VIII, 228 pages. 1992.

Vol. 393: B. Lucke, Price Stabilization on World Agricultural Markets. XI, 274 pages. 1992.

Vol. 394: Y.-J. Lai, C.-L. Hwang, Fuzzy Mathematical Programming. XIII, 301 pages. 1992.

Vol. 395: G. Haag, U. Mueller, K. G. Troitzsch (Eds.), Economic Evolution and Demographic Change. XVI, 409 pages. 1992.

Vol. 396: R. V. V. Vidal (Ed.), Applied Simulated Annealing. VIII, 358 pages. 1992.

Vol. 397: J. Wessels, A. P. Wierzbicki (Eds.), User-Oriented Methodology and Techniques of Decision Analysis and Support. Proceedings, 1991. XII, 295 pages. 1993.

Vol. 398: J.-P. Urbain, Exogeneity in Error Correction Models. XI, 189 pages. 1993.

Vol. 399: F. Gori, L. Geronazzo, M. Galeotti (Eds.), Nonlinear Dynamics in Economics and Social Sciences. Proceedings, 1991. VIII, 367 pages. 1993.

Vol. 400: H. Tanizaki, Nonlinear Filters. XII, 203 pages. 1993.

Vol. 401: K. Mosler, M. Scarsini, Stochastic Orders and Applications. V, 379 pages. 1993.

Vol. 402: A. van den Elzen, Adjustment Processes for Exchange Economies and Noncooperative Games. VII, 146 pages. 1993.

Vol. 403: G. Brennscheidt, Predictive Behavior. VI, 227 pages. 1993.

Vol. 404: Y.-J. Lai, Ch.-L. Hwang, Fuzzy Multiple Objective Decision Making. XIV, 475 pages. 1994.

Vol. 405: S. Komlósi, T. Rapcsák, S. Schaible (Eds.), Generalized Convexity. Proceedings, 1992. VIII, 404 pages. 1994.

Vol. 406: N. M. Hung, N. V. Quyen, Dynamic Timing Decisions Under Uncertainty. X, 194 pages. 1994.

Vol. 407: M. Ooms, Empirical Vector Autoregressive Modeling. XIII, 380 pages. 1994.

Vol. 408: K. Haase, Lotsizing and Scheduling for Production Planning. VIII, 118 pages. 1994.

Vol. 409: A. Sprecher, Resource-Constrained Project Scheduling. XII, 142 pages. 1994.

Vol. 410: R. Winkelmann, Count Data Models. XI, 213 pages. 1994.

Vol. 411: S. Dauzère-Péres, J.-B. Lasserre, An Integrated Approach in Production Planning and Scheduling. XVI, 137 pages. 1994.

Vol. 412: B. Kuon, Two-Person Bargaining Experiments with Incomplete Information. IX, 293 pages. 1994.

Vol. 413: R. Fiorito (Ed.), Inventory, Business Cycles and Monetary Transmission. VI, 287 pages. 1994.

Vol. 414: Y. Crama, A. Oerlemans, F. Spieksma, Production Planning in Automated Manufacturing. X, 210 pages. 1994.

Vol. 415: P. C. Nicola, Imperfect General Equilibrium. XI, 167 pages. 1994.

Vol. 416: H. S. J. Cesar, Control and Game Models of the Greenhouse Effect. XI, 225 pages. 1994.

Vol. 417: B. Ran, D. E. Boyce, Dynamic Urban Transportation Network Models. XV, 391 pages. 1994.

Vol. 418: P. Bogetoft, Non-Cooperative Planning Theory. XI, 309 pages. 1994.

Vol. 419: T. Maruyama, W. Takahashi (Eds.), Nonlinear and Convex Analysis in Economic Theory. VIII, 306 pages. 1995.

Vol. 420: M. Peeters, Time-To-Build. Interrelated Investment and Labour Demand Modelling. With Applications to Six OECD Countries. IX, 204 pages. 1995.

Vol. 421: C. Dang, Triangulations and Simplicial Methods. IX, 196 pages. 1995.

Vol. 422: D. S. Bridges, G. B. Mehta, Representations of Preference Orderings. X, 165 pages. 1995.

Vol. 423: K. Marti, P. Kall (Eds.), Stochastic Programming. Numerical Techniques and Engineering Applications. VIII, 351 pages. 1995.

Vol. 424: G. A. Heuer, U. Leopold-Wildburger, Silverman's Game. X, 283 pages. 1995.

Vol. 425: J. Kohlas, P.-A. Monney, A Mathematical Theory of Hints. XIII, 419 pages, 1995.

Vol. 426: B. Finkenstädt, Nonlinear Dynamics in Economics. IX, 156 pages. 1995.

Vol. 427: F. W. van Tongeren, Microsimulation Modelling of the Corporate Firm. XVII, 275 pages. 1995.

Vol. 428: A. A. Powell, Ch. W. Murphy, Inside a Modern Macroeconometric Model. XVIII, 424 pages. 1995.

Vol. 429: R. Durier, C. Michelot, Recent Developments in Optimization. VIII, 356 pages. 1995.

Vol. 430: J. R. Daduna, I. Branco, J. M. Pinto Paixão (Eds.), Computer-Aided Transit Scheduling. XIV, 374 pages. 1995.

Vol. 431: A. Aulin, Causal and Stochastic Elements in Business Cycles. XI, 116 pages. 1996.

Vol. 432: M. Tamiz (Ed.), Multi-Objective Programming and Goal Programming. VI, 359 pages. 1996.

Vol. 433: J. Menon, Exchange Rates and Prices. XIV, 313 pages. 1996.

Vol. 434: M. W. J. Blok, Dynamic Models of the Firm. VII, 193 pages. 1996.

Vol. 435: L. Chen, Interest Rate Dynamics, Derivatives Pricing, and Risk Management. XII, 149 pages. 1996.

Vol. 436: M. Klemisch-Ahlert, Bargaining in Economic and Ethical Environments. IX, 155 pages. 1996.

Vol. 437: C. Jordan, Batching and Scheduling. IX, 178 pages. 1996.

Vol. 438: A. Villar, General Equilibrium with Increasing Returns. XIII, 164 pages. 1996.

Vol. 439: M. Zenner, Learning to Become Rational. VII, 201 pages. 1996.

Vol. 440: W. Ryll, Litigation and Settlement in a Game with Incomplete Information. VIII, 174 pages. 1996.

Vol. 441: H. Dawid, Adaptive Learning by Genetic Algorithms. IX, 166 pages.1996.

Vol. 442: L. Corchón, Theories of Imperfectly Competitive Markets. XIII, 163 pages. 1996.

Vol. 443: G. Lang, On Overlapping Generations Models with Productive Capital. X, 98 pages. 1996.

Vol. 444: S. Jørgensen, G. Zaccour (Eds.), Dynamic Competitive Analysis in Marketing. X, 285 pages. 1996.

Vol. 445: A. H. Christer, S. Osaki, L. C. Thomas (Eds.), Stochastic Modelling in Innovative Manufactoring. X, 361 pages. 1997.

Vol. 446: G. Dhaene, Encompassing. X, 160 pages. 1997.

Vol. 447: A. Artale, Rings in Auctions. X, 172 pages. 1997.

Vol. 448: G. Fandel, T. Gal (Eds.), Multiple Criteria Decision Making. XII, 678 pages. 1997.

Vol. 449: F. Fang, M. Sanglier (Eds.), Complexity and Self-Organization in Social and Economic Systems. IX, 317 pages, 1997.

Vol. 450: P. M. Pardalos, D. W. Hearn, W. W. Hager, (Eds.), Network Optimization. VIII, 485 pages, 1997.

Vol. 451: M. Salge, Rational Bubbles. Theoretical Basis, Economic Relevance, and Empirical Evidence with a Special Emphasis on the German Stock Market.IX, 265 pages. 1997.

Vol. 452: P. Gritzmann, R. Horst, E. Sachs, R. Tichatschke (Eds.), Recent Advances in Optimization. VIII, 379 pages. 1997.

Vol. 453: A. S. Tangian, J. Gruber (Eds.), Constructing Scalar-Valued Objective Functions. VIII, 298 pages. 1997.

Vol. 454: H.-M. Krolzig, Markov-Switching Vector Autoregressions. XIV, 358 pages. 1997.

Vol. 455: R. Caballero, F. Ruiz, R. E. Steuer (Eds.), Advances in Multiple Objective and Goal Programming. VIII, 391 pages. 1997.

Vol. 456: R. Conte, R. Hegselmann, P. Terna (Eds.), Simulating Social Phenomena. VIII, 536 pages. 1997.

Vol. 457: C. Hsu, Volume and the Nonlinear Dynamics of Stock Returns. VIII, 133 pages. 1998.

Vol. 458: K. Marti, P. Kall (Eds.), Stochastic Programming Methods and Technical Applications. X, 437 pages. 1998.

Vol. 459: H. K. Ryu, D. J. Slottje, Measuring Trends in U.S. Income Inequality. XI, 195 pages. 1998.

Vol. 460: B. Fleischmann, J. A. E. E. van Nunen, M. G. Speranza, P. Stähly, Advances in Distribution Logistic. XI, 535 pages. 1998.

Vol. 461: U. Schmidt, Axiomatic Utility Theory under Risk. XV, 201 pages. 1998.

Vol. 462: L. von Auer, Dynamic Preferences, Choise Mechanisms, and Welfare. XII, 226 pages. 1998.

Vol. 463: G. Abraham-Frois (Ed.), Non-Linear Dynamics and Endogenous Cycles. VI, 204 pages. 1998.

Vol. 464: A. Aulin, The Impact of Science on Economic Growth and its Cycles. IX, 204 pages. 1998.

Vol. 465: T. J. Stewart, R. C. van den Honert (Eds.), Trends in Multicriteria Decision Making. X, 448 pages. 1998.